November 8, 2002

Fred,

I hope you enjoy reading this book as much as I enjoyed writing it.

[illegible]

AF704368

BADGES *of the* BRAVEST

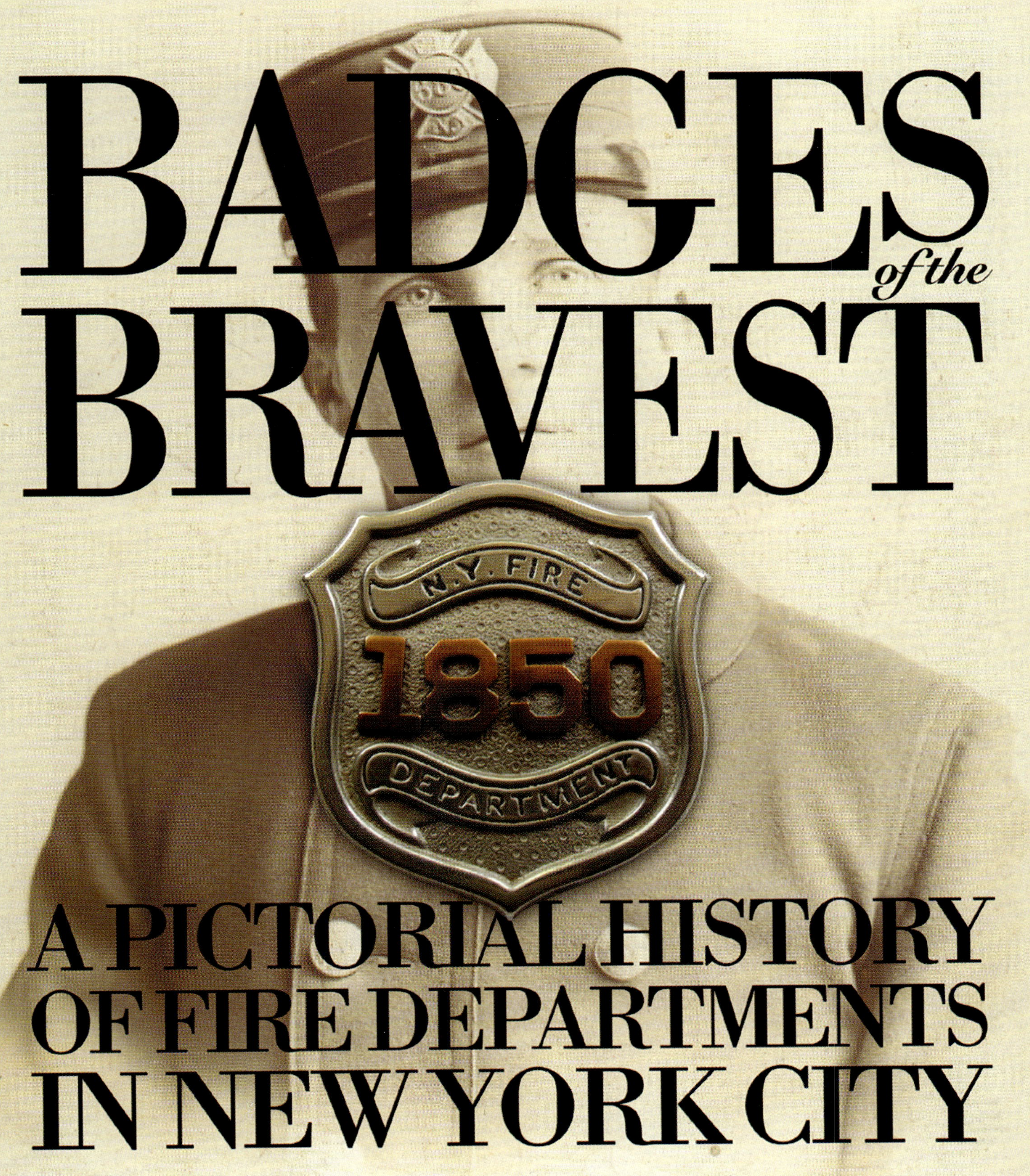

A PICTORIAL HISTORY OF FIRE DEPARTMENTS IN NEW YORK CITY

Gary R. Urbanowicz

Turner Publishing Company
Publishers of America's History
P.O. Box 3101
Paducah, Kentucky 42002-3101

Co-published by
Mark A. Thompson, Associate Publisher

For book publishing write to:
M.T. Publishing Company, Inc.
P.O. Box 6802
Evansville, Indiana 47719-6802

Pre-Press work by M.T. Publishing Company, Inc.
Graphic Designer: Elizabeth A. Dennis
Cover Design: Pauline Chu and Bill Collins

Copyright © 2002
Gary Urbanowicz

This book or any part thereof may not be reproduced without the written consent of Gary Urbanowicz and the Publishers.

The materials were compiled and produced using available information; Turner Publishing Company, M.T. Publishing Company, Inc., and Gary Urbanowicz regret they cannot assume liability for errors or omissions.

Library of Congress
Control Number 2002102263

ISBN: 1-56311-797-5

Printed in the United States of America

Limited Edition of 1000 copies of which
this book is number

____

Contents

Dedication

Dedicated to my father,
Fireman First Grade Roman P. Urbanowicz,
who wore FDNY badge number 7065 from
June 16, 1953 through July 23, 1975.

Acknowledgment

It is not possible to seek out, research, and photograph the 900+ badges that are contained within this book without a lot of help. Fortunately, I have had the cooperation of several authoritative resources to help with my investigations.

From the Department, I would like to thank Lourdes Sinistera, Administrative Assistant to the Commissioner; Lenore Koehler, Special Projects Director; Chief John J. Francese, Chief of Uniformed Personnel; Firefighter Andy Alongi of the Badge Desk; Lt. Richard Smiouskas of the Photo Unit; Acting Lt. Daniel Maye, Department Librarian, Lt. Paul Hudak, and Hon. Chief of Department Jack Lerch, Mand Library. This book could not have been written without the help of the staff of the NYC Fire Museum, Geoff Giglierano, Director and Peter Rothenberg, Collection Manager. I would also like to recognize the many New York volunteers from years gone by who retired to the Fireman's Home in Hudson, New York where their possessions, including badges, are housed and displayed in the American Museum of Firefighting. I am grateful to the Board of Trustees and especially Trustee R. Dennis Randall, Curator of the Museum and his very knowledgeable assistant, Karen Del Principe who gave much of her time and assistance.

It is amazing where NYC fire-related badges are found. Several museums outside the City made their collections available to me including Fireman's Hall in Philadelphia, thanks to Harry McKee, Curator and Dan Kenny, Assistant Curator. The best kept secret among fire museums is the Connecticut Fire Historical Society Museum, where Wayne Crossman, Dave Velapoldi and Gary Pinkham spent a hot Sunday afternoon moving and dismantling hundred-plus-pound display cases for me. Thanks for the sweat guys!

The resources of several agencies were made available to me, including permission to photograph the badges issued by them. Thanks go to Arthur Smith, Secretary of the New York Board of Fire Underwriters and Superintendent of the New York Fire Patrol; Chief Robert Henesy, Metro-North Fire Department; Michael Hurley (World Trade Center), Police Officer Lawrence Rand (LaGuardia Airport) and Raymond Brown (Bridges & Tunnels) of the Port Authority of New York and New Jersey; John Fiumano, Fire Safety Manager, National Railroad Passenger Corporation; Ms. Laura Rosen, MTA Bridges and Tunnels; Mr. Robert Martella, United Nations Safety and Security; Chief Louis Montalto, United States Coast Guard Fire Department, Governor's Island; and Ms. Margaret Czanowicki, United States Merchant Marine Academy Museum.

I had the good fortune to be able to contact Dr. Robert Holzman, the author of a landmark book on the fire service, *The Romance of Firefighting*. Dr. Holzman provided valuable input to this work based on his own research as well as his many years as a Buff of FDNY Engine Company 56. Sadly, Dr. Holzman passed away at the age of 90, prior to the completion of this book. On the few occasions that I visited him at his home, he seemed to enjoy talking about his old Buffing days as much as I enjoyed listening to him. I had the honor of calling him my friend.

A great deal of assistance was generously given to me by fellow collectors and Buffs. I would like to acknowledge and thank those who shared their collections with me:

S.F.A.D. Bruce Brenner (FDNY); Forest Hills, New York

Hon. Dep. Chief Ken Bronstein (FDNY); New York, New York

F.F. William Collins (Los Angeles County FD - Ret.); Tustin, California

Howard R. Crouch; Fairfax, Virginia

Lt. John Cunningham (FDNY-Ret.); Brooklyn, New York

F.F. Raymond Darida (FDNY-Ret.); Northfield, Connecticut

Chief Thomas D'Arcangelo (Central Islip FD); Central Islip, New York

B.C. George Eysser (FDNY); Syosset, New York

F.F. Eugene Fox (FDNY); Monsey, New York

Chief August "Chip" Greiner (Leonia P.D.); Leonia, New Jersey

Acting Capt. Robert Henesy (FDNY-Ret.); New York, New York
Lt. Paul Hudak (FDNY); Randalls Island, New York
Jacques Noel Jacobsen; Staten Island, New York
Ralph Jennings; Fort Washington, Pennsylvania
Robert and Marilyn Katzman; San Francisco, California
Hon. Dep. Cmsr. Philip Kaufman (FDNY); New York, New York
Lt. Peter Keller (FDNY-Ret.); Holbrook, New York
Hon. Medical Officer Marc Kramer, Ph.D. (FDNY); New York, New York
Hon. Chief of Dept. Jack Lerch (FDNY); New York, New York
Lt. Ernest Leves, (PD-Ret.); New Jersey
Michael Mann; Rensselaer, New York
B.C. Robert Mastrella (FDNY-Ret.); Middle Village, New York
Michael McDowell; Oakland, California
F.F. John McGuire (Hanover Park FD); Hanover Park, Illinois
F.F. Howard Meile III (Baltimore FD); Baltimore, Maryland
F.F. Arnold Merkitch (FDNY-Ret.); West Islip, New York
Brian Merlis; Lynbrook, New York
Kathleen Walsh Packard; Now, Washington
Robert Paz (NYPD-Ret.); Elmhurst, New York
James Piatti; Ringwood, New Jersey
Gary Pinkham; Colchester, Connecticut
Ira Rubin; Bayonne, New Jersey
Lt. Caesar (Sandy) Sansevero (FDNY-Ret.); Brooklyn, New York
Hon. Dep. Chief Steven Scher (FDNY); Bronx, New York
Lt. Ed Sere (FDNY-Ret.); South Ozone Park, New York
Engr. Steve Shaw (San Bernardino FD); Highland, California
Engr. Dennis Shell (Orange County Fire Authority); Downey, California
Timothy Stillman; Stonington, Connecticut
Robert Stonehill; Cortlandt Manor, New York
J. J. Sullivan; Brooklyn, New York
Terry Sullivan; Yonkers, New York
Capt. Elizabeth Tickner (Wilmington FD); Wilmington, Delaware
Lt. William White (NYC-EMS-Ret.); NewYork, New York
William T. Willis; Gradyville, Pennsylvania
Division Chief Steve Wells (Rialto FD); Rialto, California
Robert Williams; Staten Island, New York

Representatives of several companies also contributed their knowledge and resources to this project and I thank them, including Mary McGlynn, C. G. Braxmar & Company; John Campagna, Brooklyn Union Gas Company; Jane Adler, Pinkerton Security and Investigative Services; Ms. Suzanne J. Wallace, Mobil Oil Company; Louisa Bann, Tiffany & Co.; and Sandy Dunigan, ASM International.

Special thanks go to someone I met while searching for badges who has become a life-long friend. Bill Collins not only contributed hundreds of badges from his collection, he and his talented wife, Pauline, provided their graphic arts skills by designing the book's cover and creating several other illustrations.

Everyone who ever writes a book appropriately thanks their spouse for their support and understanding during the process. However, I thank my wife, Nancy, not only for her encouragement and support related to this book but also for her unending tolerance of my hobby. She has generously bought me helmets, badges, and useless paraphernalia when she wanted to buy me ties, underwear and other necessities. Throughout the world, she has accompanied me to countless fire museums when she would rather be admiring world-famous artwork and relics. And she has been gracious in the company of other fire collectors as the expression, *"The only difference between men and boys is the price of their toys,"* echoes over and over in her mind. All because she loves me...and I love her.

Foreword

The City of New York

FIRE DEPARTMENT
9 MetroTech Center   Brooklyn, N.Y. 11201-3857

Room 8W-5

THOMAS VON ESSEN
Fire Commissioner

When I was appointed Commissioner, I was presented with three badges of my choosing to carry during my tenure. At that time, I wondered both about how these badge styles came to be and about the list of my predecessors that the badge of Commissioner represents.

You hold in your hands, the first comprehensive history of the symbol of the men and women of the Fire Department of the City of New York - their badges. But, as you will see, this is not merely a catalog of these simple, yet attractive, medals. This is the story of the brave firefighters who wear them and the history of fire fighting in New York City that they reflect.

The growth and evolution of the FDNY occurred along with that of the City. In every aspect of the City's history one can see its impact on the fire protection extended by the FDNY, as well as by several specialized fire brigades and services. Prior to 1865, fire protection was provided by a volunteer group of citizens. For them, the Department was as important a social and cultural outlet, as it was a community service. Several volunteer departments continue providing fire protection in their communities to this day. What you are about to experience is a true memorial to the dedication of every individual involved in the history of fire fighting in New York City.

I am proud to be a part of this unique fraternity and am truly honored to be the Commissioner. I am proud of every uniformed firefighter who puts their life on the line everyday so that others might be safe. I am proud of every non-uniformed member who supports and complements the firefighters' work through fire prevention, education, emergency medical services, code enforcement, and administrative duties. I am proud of every New Yorker who, in the past, present and future, accepts the responsibility to protect our City with pride and professionalism. And, finally, I am proud to wear a badge of the FDNY.

Thomas Von Essen

Thomas Von Essen
Commissioner

The City of New York
FIRE DEPARTMENT
9 MetroTech Center Brooklyn, N.Y. 11201-3857

DANIEL A. NIGRO
Chief of Department
Office of the Chief of Department

Room 7W-4

The uniformed force of the FDNY has a long history of which to be proud. Since 1855 each and every one of its members has been issued a badge; as much a symbol of pride as it is a means of identification. Each is a badge of honor, courage, respect, and professionalism. For over 135 years, the Maltese cross has been recognized around the world as representing the guardians of our city who protect its multitude of communities and neighborhoods from the ravages of mankind's oldest natural enemy – fire.

Our city began as a small settlement of the 17th century and grew to become the heartbeat of the world that it is today. Our department started from bucket brigades and is now the best-equipped fire department in the world. What began with an untrained group of concerned citizens evolved into the best-trained, most revered firefighting force of dedicated professionals. Firefighters have had the responsibility to protect the lives and properties of this City since 1753 and have served it courageously whether as a member of a volunteer corps, a corporate brigade, or a civil servant in this Department whose uniform I proudly wear.

This book brings to light many of the individuals and organizations that are all a part of the firefighting history of this City, some of which have been overlooked in other historical discourses. These include our close cousins in the New York Fire Patrol and the fire insurance industry, our line organizations that promote the fraternity of firefighting, our civilian support staff without whom we could not function, our honorary officers who have contributed through the years to the well-being and recognition of our members and buffs who provide emotional and much-appreciated supportive services at the scene and in our firehouses.

With this book I look upon my own badge as the embodiment of the legacy that has been passed on to me. So, as you read through this chronicle of badges, always keep in mind the fact that each one is only meaningful when the person who wore it is thought of. For without them, these badges are mere pieces of metal.

Daniel A. Nigro
Chief of Department

In Memoriam

Between January 1, 1865 and September 10, 2001, 778 NYC firefighters gave their lives in the line of duty. Then, in a single day, nearly half that number were lost. On September 11, 2001, after this book was written but before it was printed, the worst tragedy in the history of New York City, the United States of America and especially the Fire Department of the City of New York, struck. Four terrorist attacks claimed the lives of thousands including 343 NYC firefighters. The names of the fallen firefighters, and their badge numbers, which will never be issued again, are presented here to honor their courage.

Lt Joseph Agnello*	11287
Lt Brian Ahern	1361
FF Eric Allen	2806
FF Richard Allen	3836
BC James Amato*	256
FF Calixto Anaya Jr.	3837
FF Jospeh Angelini	2630
FF Joseph Angelini Jr.	10235
FF Faustino Apostol Jr.	7371
FF David Arce	736
FF Louis Arena	1460
FF Carl Asaro	7963
Lt Gregg Atlas	1747
FF Gerald Atwood	10649
FF Gerard Baptiste	782
AC Gerard Barbara	7
FF Matthew Barnes	13088
FF Arthur Barry	8789
Lt Steven Bates	29
FF Carl Bedigian	11367
FF Stephen Belson	3942
FF John Bergin	6359
FF Paul Beyer	9120
FF Peter Bielfeld	1699
FF Brian Bilcher	10071
FF Carl Bini	13140
FF Christopher Blackwell	342
FF Michael Bocchino	3755
FF Frank Bonomo	6422
FF Gary Box	809
FF Michael Boyle	528
FF Kevin Bracken	9393
FF Michael Brennan	2622
FF Peter Brennan	2789
Capt Daniel Brethel	1204
Capt Patrick Brown	814
FF Andrew Brunn	3893
Capt Vincent Brunton	388
FM Ronald Bucca	317
FF Greg Buck	3160
Capt William Burke Jr.	91
AC Donald Burns	13
FF John Burnside	11486
FF Thomas Butler	2628
FF Patrick Byrne	8883
FF George Cain	11675
FF Salvatore Calabro	7548
Capt Frank Callahan	141
FF Michael Cammarata	1138
FF Brian Cannizzaro	11126
FF Dennis Carey	742
FF Michael Carlo	5592
FF Michael Carroll	563
FF Peter Carroll	7562
FF Thomas Casoria	6259
FF Michael Cawley	1614
FF Vernon Cherry	633
FF Nicholas Chiofalo	9305
FF John Chipura	10830
FF Michael Clarke	12716
FF Steven Coakley	156
FF Tarel Coleman	2411
FF John Collins	12234
FF Robert Cordice	13423
FF Ruben Correa	11021
FF James Coyle	3479
FF Robert Crawford	11904
Lt John Crisci	3039
DC Dennis Cross*	102
FF Thomas Cullen 3rd	1694
FF Robert Curatolo	4263
Lt Edward Datri	1125
FF Michael Dauria	3928
FF Scott Davidson	8418
FF Edward Day	11189
BC Thomas DeAngelis	23
Lt Manuel Delvalle*	3925
FF Martin Demeo	4530
FF David DeRubbio	12769
Lt Andrew Desperito	511
BC Dennis Devlin	141
FF Gerard Dewan	1810
FF George DiPasquale	5709
Lt Kevin Donnelly	3231
Lt Kevin Dowdell	927
DC Raymond Downey*	197
FF Gerard Duffy	11755
Capt Martin Egan Jr.	743
FF Michael Elferis	10325
FF Francis Esposito	1636
Capt Michael Esposito*	3045
FF Robert Evans	6240
BC John Fanning	532
Capt Thomas Farino	158
FF Terrence Farrell	2971
BC Joseph Farrelly*	90
1st Dep Cmsr William Feehan	—
FF Lee Fehling	6799
FF Alan Feinberg	603
FF Michael Fiore	3727
Capt John Fischer*	283
FM Andre Fletcher*	8516
FF John Florio	213
Lt Michael Fodor	470
FF Thomas Foley	10347
Lt David Fontana*	1627
FF Robert Foti	11202
Lt Andrew Fredericks*	217
Lt Peter Freund	1548
FF Thomas Gambino Jr.	7223
Chief of Dept Peter Ganci	—
Lt Charles Garbarini	197
FF Thomas Gardner	6489
FF Matthew Garvey	2421
FF Bruce Gary	2523
FF Gary Geidel	1915
BC Edward Geraghty	132
FF Denis Germain	8758
Capt Vincent Giammona*	1331
FF James Giberson	804
Lt Ronnie Gies*	11524
FF Paul Gill	10380
Lt John Ginley	726
FF Jeffrey Giordano	8062
FF John Giordano (1)	2803
FF Keith Glascoe	10382
FF James Gray	762
BC Joseph Grzelak	279
FF Jose Guadalupe	10445
Lt Geoffrey Guja	3205
Lt Joseph Gullickson	1003
Lt David Halderman*	10652
Lt Vincent Halloran	1440
FF Robert Hamilton	1063
FF Sean Hanley	9137
FF Thomas Hannafin	10467
FF Dana Hannon	2185
FF Daniel Harlin	8045
Lt Harvey Harrell	157
Lt Stephen Harrell	533
BC Thomas Haskell Jr.	250
FF Timothy Haskell	12885
Capt Terence Hatton	463
FF Michael Haub	1695
Lt Michael Healey	1004
FF John Heffernan	11963
FF Ronnie Henderson	10728
FF Joseph Henry	3538
FF William Henry	1959
FF Thomas Hetzel	4938

Name	Badge
BC Brian Hickey*	1217
Lt Timothy Higgins	903
Ff Jonathan Hohmann	17
FF Thomas Holohan	5769
FF Joseph Hunter	5485
Capt Walter Hynes	139
FF Jonathan Ielpi	12642
Capt Frederick Ill Jr.	714
FF William Johnston	1246
FF Andrew Jordan	5894
FF Karl Joseph	3585
Lt Anthony Jovic	982
FF Angel Juarbe Jr.	6622
Chaplain Mychal Judge	—
FM Vincent Kane*	341
DC Charles Kasper*	31
FF Paul Keating	5064
FF Richard Kelly Jr.	8650
FF Thomas R. Kelly	11408
Lt Thomas W. Kelly*	1964
FF Thomas Kennedy	12984
Lt Ronald Kerwin	385
FF Michael Kiefer	3597
FF Robert King Jr.	4316
FF Scott Kopytko	10587
FF William Krukowski	10600
FM Kenneth Kumpel*	11335
FF Thomas Kuveikis	1333
FF David LaForge	3728
FF William Lake	1571
FF Robert Lane	10609
FF Peter Langone	9286
FF Scott Larsen	432
Lt Joseph Leavey	1398
FF Neil Leavy	12911
FF Daniel Libretti	10294
PAR Carlos Lillo	3723
FF Robert Linnane	1300
Lt Michael F. Lynch*	2315
FF Michael F. Lynch	12098
FF Michael Lyons	6946
Lt Patrick Lyons*	524
FF Joseph Mafeo	9621
FF William Mahoney	1160
FF Joseph Maloney	740
DC Joseph Marchbanks Jr.*	651
Lt Charles Margiotta	3061
FF Kenneth Marino	767
FF John Marshall	1750
Lt Peter Martin	211
Lt Paul Martini	1373
FF Joseph Mascali	8770
FF Keithroy Maynard	1757
FF Brian McAleese	8264
FF John McAvoy	4953
FF Thomas McCann	9079
Capt William McGinn*	3052
BC William McGovern	37
FF Dennis McHugh	12991
FF Robert McMahon	847
FF Robert McPadden	2397
FF Terence McShane	1409
FF Timothy McSweeney	8141
FF Martin McWilliams	7015
FF Raymond Meisenheimer	4745
FF Charles Mendez	7386
FF Steve Mercado	11632
FF Douglas Miller	10401
FF Henry Miller Jr.	9650
FF Robert Minara	6034
FF Thomas Mingione	266
Lt Paul Mitchell	8815
BC Louis Modafferi*	13
Lt Dennis Mojica	810
FF Manuel Mojica	9017
FF Carl Molinaro	8437
FF Michael Montesi	12492
Capt Thomas Moody	1489
BC John Moran	57
FF Vincent Morello	1447
FF Chirstopher Mozzillo	1908
FF Richard Muldowney Jr.	9312
FF Michael Mullan	7830
FF Dennis Mulligan	6066
Lt Raymond Murphy	742
Lt Robert Nagel	1072
Lt John Napolitano*	3709
FF Peter Nelson	9798
FF Gerard Nevins	3181
FF Dennis Oberg	2939
Lt Daniel O'Callaghan	3257
FF Douglas Oelschlager	2456
FF Joseph Ogren	13075
Lt Thomas O'Hagan	234
FF Samuel Oitice	3190
Capt William O'Keefe	749
FF Patrick O'Keefe	10548
FF Eric Olsen	1524
FF Jeffrey Olsen	8984
FF Steven Olson	2205
FF Kevin O'Rouke	5837
FF Michael Otten	1653
FF Jeffrey Palazzo	8820
DC Orio Palmer*	253
FF Frank Palombo	10871
FM Paul Pansini*	2206
DC John Paolillo*	503
FF James Pappageorge	3322
FF Robert Parro	7403
FF Durrell Pearsall	12338
Lt Glenn Perry	438
Lt Philip Petti	427
Lt Kevin Pfeifer	878
Lt Kenneth Phelan	669
FF Christopher Pickford	1625
FF Shawn Powell	1876
FF Vincent Princiotta	11014
FF Kevin Prior	3918
BC Richard Prunty	131
FF Lincoln Quappe	12879
Lt Michael Quilty	459
PAR Richardo Quinn	1736
FF Leonard Ragalia	12801
FF Michael Ragusa	2981
FF Edward Rall	4842
FF Adam Rand	4126
FF Donald Regan	2485
Lt Robert Regan	248
FF Christian Regenhard	d2986
FF Kevin Reilly	1907
Capt Vernon Richard*	218
FF James Riches	734
FF Joseph Rivelli Jr.	2489
FF Michael E. Roberts	6611
FF Michael E. Roberts	13392
FF Anthony Rodriquez	4095
FF Matthew Rogan	2478
FF Nicholas Rossomando	12211
FF Paul Ruback	12549
FF Stephen Russell	13124
Lt Michael Russo	300
BC Matthew Ryan	170
FF Thomas Sabella	12896
FF Christopher Santora	3041
FF John Santore	12127
FF Gregory Saucedo	6754
FF Dennis Scauso	10991
FF John Schardt	5088
BC Fred Scheffold	633
FF Thomas Schoales	619
FF Gerard Schrang	11838
FF Daniel Suhr	5717
FF Gregory Sikorsky	6786
FF Stephen Siller	3870
FF Stanley Smagala Jr.	979
FF Kevin Smith	6504
FF Leon Smith Jr.	12834
FF Robert Spear Jr.	3090
FF Joseph Spor	4957
BC Lawrence Stack	434
Capt Timothy Stackpole	818
FF Gregory Stajk	1342
FF Jeffrey Stark	10499
FF Benjamin Suarez	10298
Lt Christopher Sullivan	1577
FF Brian Sweeney	12177
FF Sean Tallon	3710
FF Allan Tarasiewicz	2563
FF Paul Tegtmeier	2817
FF John Tierney	3135
FF John Tipping 2nd	8572
FF Hector Tirado Jr.	3714
FF Richard Vanhine	252
FF Peter Vega	9700
FF Lawrence Veling	7932
FF John Vigiano 2nd	3436
FF Sergio Villanueva	2852
FF Lawrence Virgilio	12500
Lt Robert Wallace	1457
Lt Jeffrey Walz*	5715
Lt Michael Warchola	681
Capt Patrick Waters	341
FF Kenneth Watson	11685
FF Michael Weinberg	7485
FF David Weiss	9489
FF Timothy Welty	10983
FF Eugene Whelan	3206
FF Edward White	13401
FF Mark Whitford	13404
Lt Glenn Wilkinson	639
BC John Williamson	318
Capt David Wooley	88
FF Raymond York	11950

* Promoted posthumously on 9/16/01 retroactive to 9/10/01

Compiled by Wadell English, FDNY Badge Desk

Introduction

"I find the law authorizing the issue of badges to members of the department to be productive of much evil. It has not accomplished what it was designed to do, but has added largely to the cause for complaint. The number of badges that are either lost or stolen amounts to over one thousand at the present time, and it is growing to be a source of dissatisfaction among members of the department at the necessity of wearing them. The ownership of a stolen badge gives free access to thieves and other dangerous persons to the close proximity of fires, where, under the recognized shield of the department they can perform their nefarious calling without suspicion or interruption.

I would advise an amendment to the law - that the badges be abolished altogether, save for the Engineer Board and officers of companies."

Chief Engineer Harry Howard
Report to the NYC Council 1855

Clearly, the Council did not heed Chief Howard's recommendation to eliminate the use of badges by the Fire Department of the City of New York. And, as they say, the rest is history.

As my interest in FDNY history and collecting fire memorabilia took a serious turn, I became aware of the many different styles and titles of badges issued by the Department. As I encountered some of the more intriguing and unexpected styles and titles, I realized that there was a need to catalog this historic aspect of the Department. When I began to categorize the different badges, I was amazed by the number of badges issued by or associated with the FDNY.

Initially, it seemed that writing a book about fire badges would be straightforward and its scope would be clear cut. However, it quickly became apparent that so many badges were in some way associated with firefighting in New York City, that the project would be much bigger than I ever anticipated. I considered narrowing the scope of the book and limiting the badges only to those issued by the Department since 1865. But I did not think justice could be done to this subject without throwing the net very wide and capturing as many badges as possible that were used within the physical borders of what forms the City of New York today. As a result, over 900 badges are shown.

All of the classic FDNY history books mention a badge here or there, but no single source served as the foundation for my research. In fact, it was very difficult to find firm documentation on many badges. The reader might be frustrated to find some desirable information lacking, particularly with regard to the dates of many of the badges. Although some sources speculated about their use or time frames, unless written documentation was found, I chose not to include the information in this book. Some badges with an FDNY designation were difficult to authenticate and, in some cases, to explain.

Assembling this book has been a very rewarding experience. I had the pleasure of meeting many people with the same interest in the FDNY as I have. All were cooperative and hospitable, often welcoming me into their homes with my camera equipment. In addition to improving my own appreciation for the history of firefighting in New York, this particular area of fire memorabilia, and my knowledge of New York City history, I came to realize one other subtlety of this topic. The badges contained in this book illustrate the diverse cultural and social history that make New York City so unique. Take for example, the Theater Detail badge. New York has been, and continues to be, the home of the *Great White Way* - Broadway, the world center of stage entertainment. Badges from the New York Port of Embarkation and the Department's own Marine Bureau, reflect New York's past role as the gateway to the United States for ocean-going travelers during war and peace. The Department's line organizations parallel the ethnic and cultural diversity of the City, and their badges or logos often incorporate symbols from their ancestry, a shamrock, an eagle, a Spanish galleon.

It is my hope that readers of this book find it interesting and useful. I imagine that many will use it as a reference for badge collecting. This was certainly my original motivation for undertaking the project. But, I also hope that readers will come to appreciate, as I did, the broader historical significance of badges that tell the story of a very colorful and exciting part of New York City history.

CHAPTER 1

A Brief History Of New York City And Its Fire Department

"The City"

To most people who live outside of the New York City metropolitan area, and even to some of those within, the *City of New York* is an enigma. This brief overview is presented to help the reader understand a little bit about the City's history, as well as its geography and related terminology.

New York State is divided into sixty-two counties. New York City comprises five of these counties, namely *New York, Kings, Queens, Richmond* and *Bronx.* However, when the Greater City of New York was consolidated in 1898, all or part of these state counties became administrative districts known as city boroughs. To confuse matters more, the borough name is not necessarily identical to the county name. The result is the following:

New York in 1850

County Name		Borough Name
New York	=	Manhattan
Kings	=	Brooklyn
Richmond	=	Staten Island
Queens	=	Queens
Bronx	=	"The" Bronx

Prior to consolidation, *New York* consisted solely of what we now call Manhattan. In fact, even though there are few, if any, remaining residents who can recall the City before 1898, many present-day residents of outlying boroughs (which are part of New York City) still use the term, *"the City"* or, *"I'm going into the City"* to mean Manhattan.

The terms *downtown, midtown,* and *uptown* refer to sections of Manhattan, but not all of it. For example, there are several communities with their own identifying monikers below downtown, and above uptown. The term downtown is also used in the borough of Brooklyn to mean the central administrative district. These terms are not used in any of the other boroughs.

There are many other names associated with New York City that add to the confusion. Most, if not all of these are neighborhood names which were villages, towns or even cities, prior to consolidation. Their names remain in use today and, in many cases, continue to be post office locations. So, to an outsider, someone might say they live in New York City, but to a fellow New Yorker they might say Greenpoint (in Brooklyn), Little Neck (in Queens), Greenwich Village (in Manhattan), New Dorp (in Staten Island), Riverdale (in the Bronx) or any one of hundreds of other neighborhoods. Significantly, when these current neighborhoods were independent towns, many of them had their own fire department. In some cases, they may have had just one or two companies, but others, like the Edgewater Fire Department in Staten Island or the Newtown Fire Department in Queens, were quite large. In all, New York City has been served by as many as 500 fire companies organized in 90 independent departments.

Two New York City boroughs (Manhattan and Staten Island) are islands, two others (Brooklyn and Queens) are a portion of an island (Long Island), and only one (The Bronx) is on the continental U.S. mainland. There are numerous small, inhabited and uninhabited islands that are a part of New York City, each of which falls within a county boundary and is considered part of a respective borough (e.g., City Island is a part of the Bronx). Governors Island, a federal property, lies in New York harbor and does not receive fire or other services from the City. Due to travel barriers created by these islands, the City grew like a wheel as access improved, with Manhattan as the hub and the other boroughs as the spokes.

There is one Mayor for the City of New York and a governing City Council. Each of the five boroughs is under the administration of an elected Borough President. There is only one City fire department, ten remaining volunteer departments, and several small, independent organizations involved in fire protection or geographically-limited fire fighting including the New York Fire Patrol, and the U.S. Coast Guard Fire Department on Governors Island. This is much less confusing than it is for police agencies which, until very recently, had several City police departments in addition to numerous federal and state law enforcement agencies operating within the city limits.

Over eight million people inhabit the environs of New York City. If the borough of Brooklyn were still a free-standing city today, as it was prior to 1898, it would be the fourth largest city in the United States by population. It has the highest population density in the country with approximately 24,000 people per square mile. In the borough of Manhattan, the World Trade Center was occupied on a daily basis by approximately 50,000 workers and an additional 50,000 visitors. That is more people in one commercial complex than inhabit many U.S. villages, towns, and cities. To put this into a firefighting perspective, in 1993 a bomb in one of the building's parking garages required its complete evacuation and resulted in six fatalities. On September 11, 2001, in the City's worst disaster, the World Trade Center was reduced to rubble and nearly 3,000 people, including 343 New York City firefighters, lost their lives.

The Fire Department

Prior to 1874, the New York Fire Department served only, what is now, the borough of Manhattan. Today, it serves all five boroughs and over eight million inhabitants. In the intervening years, several fire departments, both paid and volunteer, served the various areas. The following is a brief history of the Fire Department of the City of New York (FDNY), much of which parallels the evolution of the badges depicted in this book.

Pre-1865

During the first 240 years since the settlement of New York as "New Amsterdam" by the Dutch in 1625, fire protection was provided by volunteers. Like many other growing communities during those centuries, fire fighting evolved from ad hoc bucket brigades to organized fire companies. As firefighting became a social as well as civic activity, we can follow many developments in the regalia associated with the fire service, and in the evolution of more effective firefighting equipment. Much of the regalia and the equipment are the coveted items of fire collectors throughout the world today. Probably the three most prominent collectibles are helmets, badges, and trumpets, not to diminish the value of other items from this period.

The first unified attempt to combat fires in New Amsterdam occurred in 1648, when Governor Peter Stuyvesant authorized the expenditure of fine money, collected for the fire hazard of a dirty chimney, to purchase firefighting buckets, hooks, and ladders. At the same time,

A New York fire engine in 1733.

the unpaid position of Fire Warden was established and four men were so appointed. It wasn't until 1657 that the first bucket brigades were organized. The following year, "The Prowlers" were hired and went on duty each night to patrol for fires. They were equipped with buckets and hooks that they used in their attempts to extinguish a blaze or stop the spread of a fire.

The first hand-drawn fire pumper arrived in New York City in 1731 and was housed in a shed behind City Hall. As more equipment was purchased and more locations were established for its storage, it became clear that an organization was needed to ensure that in case of a fire, the equipment would be taken from storage, brought to the scene, and operated by a group of able-bodied citizens. For this purpose, the General Assembly passed an Act in December 1737 to establish the "Volunteer Fire Department of the City of New York." A reorganization occurred in 1798 and the department was incorporated officially as the "Fire Department of the City of New York" (FDNY). It is by this name and acronym that the Department is currently known.

Coat of Arms of the volunteer New York Fire Department.

The use of metal badges as a symbol of membership in the FDNY began some time during the early volunteer era of the 19th century. Individual companies, who were known for their intense rivalries, instituted badges as a symbol of membership. In 1855, the City authorized the development of one badge to be displayed by all members of the New York Fire Department. It was a simple, shield-shaped badge (roughly the size and shape of a current NYC Police Officer's badge) with an upper banner displaying "NY Fire," a lower banner displaying "Department," and applied numbers corresponding to the member's number on the City roster of firemen. The Act also established the significance of possessing an FDNY badge. It directed police officers to form fire lines at every fire and not, "under any circumstances permit any person to pass unless he showed the badge or wore the uniform of the Insurance Patrol."

In 1860, the shield-shaped badge was replaced by three badges representing the three types of companies in the Department at that time, Engine, Ladder, and Hose. Each

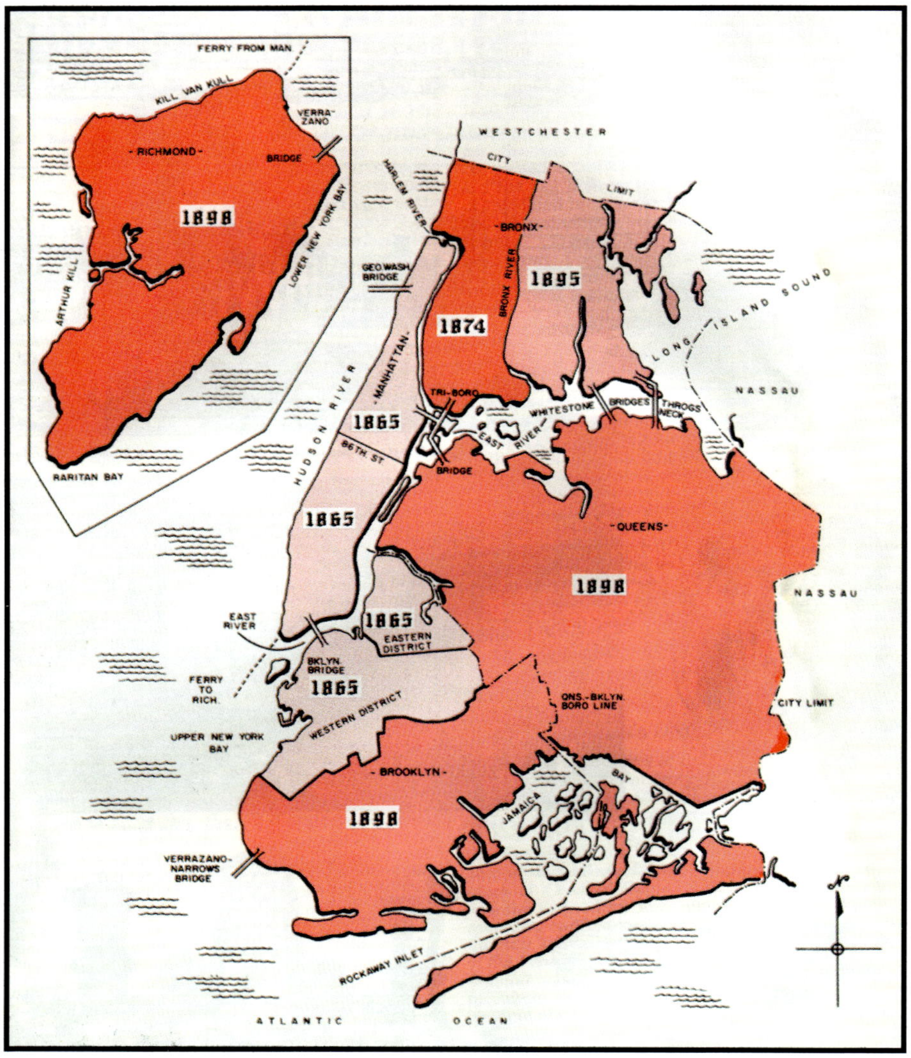

Chronological map showing the growth of the City of New York.

had a distinctive shape and a raised image of the apparatus particular to the type of company (hand-pump engine, ladder truck, or hose cart) with the company number in large applied numerals and the member's number in a smaller, engraved fashion. These badges remained in use until the organization of the Metropolitan Fire Department in 1865.

1865 to 1898

In 1865, Section 23, Chapter 243 of the Laws of 1865 (New York State) was enacted and entitled, "An Act to Create a Metropolitan Fire District and Establish a Fire Department Therein." The Act placed control of the Department in the hands of the Governor. It formed the basis of today's paid department and included within the "Fire District" the fire departments of Brooklyn. However, Brooklyn was protected by a multitude of volunteer companies organized into Eastern and Western Districts and became part of the Metropolitan Fire District (MFD) in theory only. In fact, it wasn't until 1869 that another Act, "An Act to Reorganize the Fire Department of the City of Brooklyn," was created to convert the volunteers in Brooklyn to a paid force.

On May 21, 1870, the City regained control of the Department and changed its name back to the "Fire Department of the City of New York," the name originally designated in 1798. The acronym, "FDNY" was put to widespread use. It was also at this time that the current Maltese cross badge was adopted for firemen.

The City's first expansion beyond Manhattan island occurred in 1874 when areas of southern Westchester County were annexed to form the Borough of the Bronx. At that time, the 10th Battalion of the FDNY was designated there, taking over the existing volunteer force. The Bronx was expanded in 1895 and 1898, progressively eliminating additional volunteer companies, although two remaining volunteer companies continue to operate today.

Prior to consolidation into the Greater City of New York, Queens residents were protected by the paid Long Island City Fire Department plus fifteen volunteer departments comprising sixty-three companies. On Staten Island, forty-seven companies, organized into eight volunteer fire departments were in operation.

Post-1898

In the years after consolidation of the City, the Department gradually took over areas previously served by volunteers. In 1905, the first volunteer companies were disbanded on Staten Island, and most others were replaced with FDNY units during the period from 1915 to 1937. Then, and to the current day, only two volunteer departments remain there. Large scale conversion to the paid Department in Queens occurred during 1912 with some volunteers lingering until 1929. Today, five volunteer companies remain in service.

In 1981, the FDNY appointed its first female members to the uniformed force. At that time, the title of Fireman was officially changed to Firefighter. These two terms are utilized in the book consistent with their historic time frames.

Over the past 135 years, innumerable changes have occurred within the Department. They have included the formation and/or discontinuation of various divisions, bureaus, and companies, as well as various uniformed and non-uniformed positions. Many of these changes are reflected in the evolution of badges through the years, as illustrated in the following chapters.

CHAPTER 2

Producing And Distributing Badges

***"Badge (baj) n. 1. an emblem worn to show rank, membership, etc. 2. any distinctive sign, etc."* from Webster's New World Dictionary of the English Language**

When the officials of the earliest fire department in New York were appointed, they were to be issued a sign of their authority. At that time in 1648, the badge of office for members of the Fire Wardens was, *"a wand with a gilded flame on the top."* (The *wand* was a staff measuring approximately fifty-four inches.) This was a very effective and visible means of identifying these men with a very important job. In 1813, the Common Council developed a standardized identification system for the various positions and ranks within the Fire Department of the City of New York. Although referred to in some documents as a badge, it was not what we think of today as a badge. It was, what we call today, the frontpiece of a fire helmet. Although the wand of the Fire Warden is long since extinct, the helmet frontpiece remains a useful means of identifying a firefighter's rank, company and company type. An entire book could be dedicated to helmet frontpieces, so this form of badge is not included.

In 1855, Police Captain J. Murray Ditchet of the Fourth Ward suggested a method to decrease the interference by crowds and "Buffs" at the scene of a fire by having the police establish a perimeter that would prohibit anyone but firemen and insurance patrolmen from entering (see Chapter 14). In order to identify authorized members of the Department, he suggested the issuance of a standardized badge. The Common Council approved this idea and implemented it in, *"An Act for the better regulation of the Firemen of the city of New York,"* passed March 29, 1855. (Captain Ditchet later became a New York City Volunteer Fireman, joining Protection Engine Company 5 in 1859.)

Specifications for the actual badge to be used were contained in a subsequent Ordinance. It read as follows:

> 1. All members of the New York Fire Department, and such exempt members as are authorized by the Commissioners of the Fire Department, shall hereafter when on duty as firemen, wear the leathern cap, as heretofore used, or a badge, as hereinafter provided.
>
> 2. The badge mentioned in the preceding section shall be made of Prince's metal, bearing the words "New York Fire Department" and each badge shall bear a distinct number in raised figures thereon, of white metal; and the badge to be worn by exempt firemen, as provided for in Section 3d., of this ordinance, shall be composed of white metal, with figures thereon of Prince's metal; but in all other respects to conform to the badge herein described for use of the active member of the Department; said badges shall be struck from dies, which shall be exclusively the property of the Corporation of the Clerk of the Common Council; and all of such badges shall be numbered, as the Commissioner of the New York Fire Department may direct.

The Board of Commissioners

OF THE NEW YORK FIRE DEPARTMENT

New York May 12 1865

HEREBY GIVE PERMISSION, To Edward Loudy residing at No. 133 E 25 St lately connected with Engine Company No. 7 to wear Badge Register No. 34 subject however to the penalties prescribed by Sections No. 7 8 & 9 of an ordinance of the Common Council passed June 27th 1859 and also to a Resolution of the Commissioners.

R. P. Abell Clerk. John J. Gorman Commissioner

Section 7. Every exempt fireman receiving a Badge shall have the permission to wear the same renewed sometime during the month of May in each year, otherwise such permission shall be deemed revoked, and the said Commissioners may, at any time, revoke any permission so granted by them, but in case such permission is not renewed, or is revoked, it shall be the duty of such exempt fireman to restore the badge previously worn by him to the said Commissioners.

Section 8. Any member of the New York Fire Department or exempt fireman, who shall loan his badge to any person, shall be liable to the penalties as set forth in section 1st of this Ordinance, and if any badge shall become broken or the figures be displaced while in the possession of such member or exempt fireman; he shall return the same forthwith to the said Commissioners, and in default of the same, said member may be expelled or suspended, and such exempt fireman shall have the permission to wear the same revoked.

Section 9. Any member of the New York Fire Department, or exempt firemen who shall violate any of the foregoing sections of this Ordinance, shall be subject to a fine of not less than twenty five nor over two hundred and fifty Dollars, and to imprisonment for the term of ten days, and all fines so collected shall be paid over to the Treasurer of the New York Fire Department Fund.

RESOLUTION OF COMMISSIONERS.

RESOLVED. That each and every exempt member of the New York Fire Department to whom permission shall be given by this Board to use a Badge, shall on any change of residence of such exempt Member give immediate notice of the same to this Board under the penalty of having the said permission revoked.

Badge certificate dated May 12, 1865 reads, "The Board of Commissioners of the New York Fire Department Hereby Give Permission to Edward Loudy residing at No. 133 East 25th Street lately connected with Engine Company No. 7 to wear Badge Register No. 34 subject however to the penalties prescribed by Sections No. 7, 8 & 9 of an ordinance of the Common Council passed June 27th 1859 and also to a Resolution of the Commissioners."

Although badges were used by the FDNY before 1855, they were not standardized, authorized, or issued by the Department itself; individual companies designed and purchased them. As much as New York prides itself in claiming many "firsts," it does not hold this distinction with regard to its fire department. We know that Cincinnati was the first city to have a fully paid fire department in 1853 and the Chicago Fire Department introduced the first sliding pole in April 1878. It has not been established which fire department in the country first issued a standardized badge to all of its members but, the Boston Fire Department issued a standardized badge as early as 1833. Neighboring Brooklyn Fire Department, which became part of the FDNY in 1898, issued a standard badge to its members prior to 1855 (see Chapter 4). It does appear though that the FDNY may have been the first department in the U.S. to adopt the Maltese cross as the symbol for firemen, a symbol now recognized worldwide.

The 1855 Act began a very interesting history of badges for all of the various positions and titles to evolve in the Department. This chapter details some of the logistics that have supported, and continue to support, the practice of issuing badges.

Badge Shapes

A wide variety of shapes have been used in the design of FDNY badges. While it would appear that they have been standardized to two since 1937, there have been periodic variations since then.

The first official badge of the Department issued in 1855 was a, "pinched-shield shape" which was common to many badges of that era as well as today. The shape was used again in 1865, although somewhat modified when the Metropolitan Fire Department was established. These were the only two uses of the pinched-shield shape within the Department itself. This shape was also used for the fire lines badge of 1900.

Original design sketch of the 1857 Engineer's badge.

In 1857, the badges for Engineers were a simple round shape. Though the simplest badge in shape, they were without a doubt the most exquisite badges ever issued by the Department. The first fire lines badge approved by the Department in 1884 was also round.

In 1860, a set of three badges were designed with distinctive shapes for the company each represented. They were square, round, and frontpiece-shaped, for ladder, hose, and engine companies, respectively. An image of the typical company apparatus also appeared in relief. The badges were designed by Samuel Burhans, Jr. who was a prominent member of the Department and President of the Coney Island and Brooklyn Railroads. He was noted by A.E. Costello to be, *"an enthusiast in all matters pertaining to the history of the Volunteer Department."*

The 1860 badges are commonly found during excavations of Civil War battle sites and sold to Civil War collectors. Many New York firemen enthusiastically enlisted in the Union Army and wore the NYFD badge on their military uniform. From the standpoint of a fire collector, many of the badges found at these sites are not desirable due to the damage caused by electrolysis which is produced when the dissimilar metals of the badge and the applied numbers are buried in certain soil. Nevertheless, the badges excavated from Civil War sites illustrate yet another interesting aspect of the Department's history.

Capt. George LeFort of the 11th Regiment, New York State Volunteer Infantry (First Fire Zouaves) wearing NYFD Engine Co. 26 badge on his military uniform. Capt. LeFort was killed in action on May 20, 1864. (Photo courtesy of Gil Barrett Collection - U.S. Army Military History Institute.)

Samuel Burhans, Jr. designer of the 1860 NYFD badges.

A Maltese cross was first introduced in 1865 as the hat badge of the Metropolitan Fire Department. The Department's Rules and Regulations specified the badges to be:

> *"...a white Maltese cross, with the appropriate emblems of the Department in the center, the letters "M.F.D.," and the number (numerically) on the points, and placed in the center of the front of the cap."*

Maltese cross cap device introduced by the Metropolitan Fire Department in 1865.

Historically, crosses have been used as organizational or personal symbols as far back as recorded history. Since crosses that have arms of equal size are so similar to one another, it cannot be said with certainty that the cross used by the fire service is patterned after the true "Maltese cross," the cross used by the Knights of St. John. The Knights utilized an eight-pointed cross, resembling four arrowheads converging in the center. In 1259, a decree by Pope Alexander IV recognized the cross as their symbol and also identified their battle surcoats of red. However, what we call the Maltese cross more closely resembles a Cross Pattée. This misuse of terminology is not limited to the fire service. For example, the Royal Warrant for the institution of England's Victoria Cross (equivalent to the U.S. Medal of Honor) specifies that it, "shall consist of a Maltese Cross of bronze," but there is no doubt that it too is a Cross Pattée.

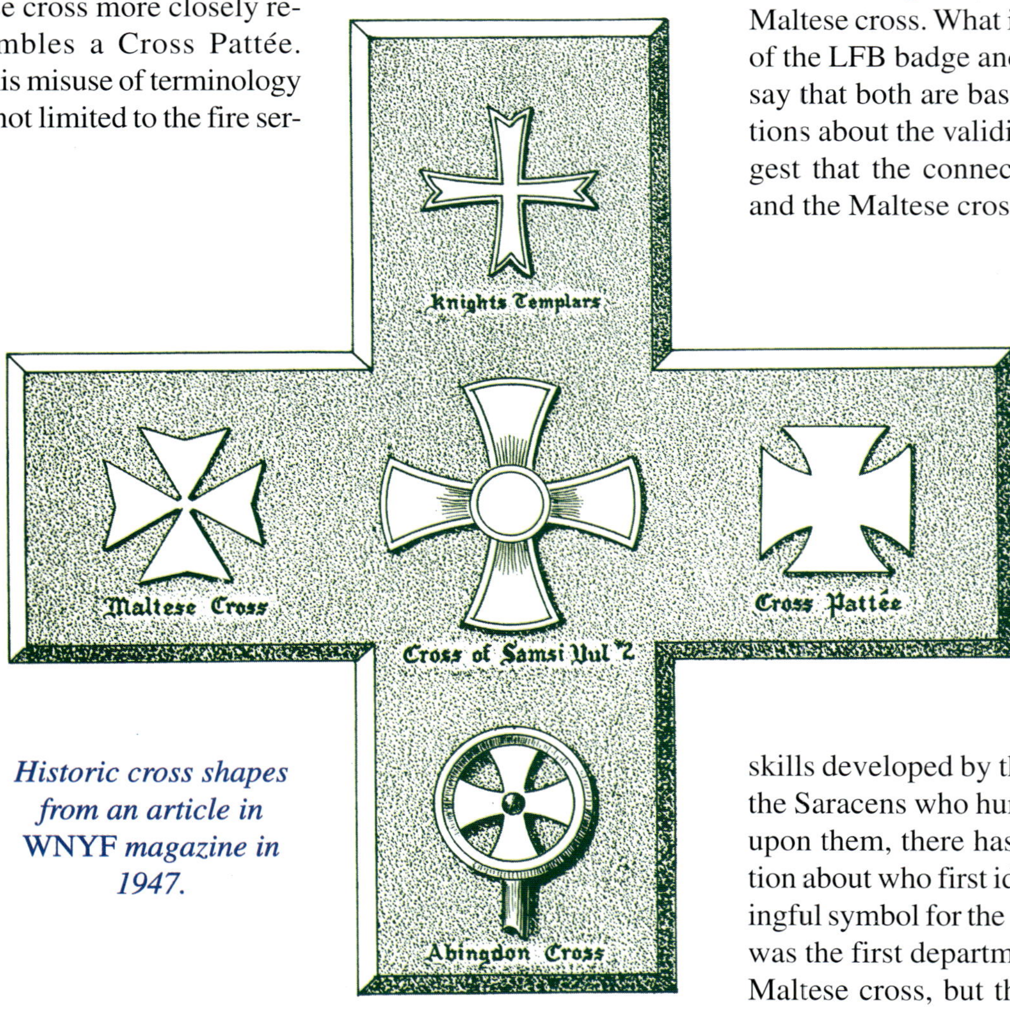

Historic cross shapes from an article in WNYF *magazine in 1947.*

The "true" Maltese cross, symbol of the Knights of St. John.

It is interesting to note the significance that the London Fire Brigade (LFB) has placed on its rendition of the Maltese cross. What is most striking is to look at the shape of the LFB badge and compare it to the FDNY badge. To say that both are based on the Maltese cross raises questions about the validity of their derivation. This may suggest that the connection between the fire service cross and the Maltese cross is based more on the similarity between the principles of the fire service and of the Knights of Malta than on any similarity in the symbols themselves.

Several fire departments in Canada do use a badge that is, without question, a Maltese cross. It is modified by a center disk displaying the city seal and department name and sunrays between the arms of the cross or arrowheads.

Although much has been written about the firefighting skills developed by the Knights of St. John when battling the Saracens who hurled glass bombs of burning naphtha upon them, there has never been any published speculation about who first identified the Maltese cross as a meaningful symbol for the fire service. It appears that the FDNY was the first department in the United States to adopt the Maltese cross, but there is no documentation on why it was chosen or who proposed its use. Several other theories, developed from the research conducted for this book,

The Eight-Pointed Star On A Firefighter's Badge

The eight-pointed star is the cross of Saint John. It originated in the Maltese Cross, the emblem of the Knights of Malta, and was used by the Knight of Saint John of Jerusalem at the time of the crusades.

The points or 'Tenets' were said to represent the knightly virtues of:

Since these represent the qualities required by a Firefighter, the star was chosen for the insignia of the national and Auxiliary Fire Service and has remained so to the Fire Service of today.

Significant of the Maltese cross from the London Fire Brigade. (Courtesy of the London Fire Brigade.)

Maltese cross badge used by Canadian fire services.

suggest that the idea to use the Maltese cross for the FDNY was not originally based on any connection to its use by the Knights of St. John.

During the Civil War, two regiments of Union troops comprised solely of New York City firemen were recruited. Officially they were the 11th and 73rd New York State Volunteer (Infantry) Regiments, but came to be known as the First and Second Fire Zouaves, respectively. Their flamboyant uniforms, copied from the French Zouave troops of northern Africa, called attention to them; unfortunately as much from enemy fire as from their admiring public. Due to heavy losses at the Battle of Bull Run, the 11th Regiment was disbanded on June 2, 1862. The 73rd fought through the end of the war. Among their many engagements, the 73rd participated in the Battle of Gettysburg where a monument of a fireman and a solider side-by-side, stands in the Peach Orchard where they fought. In 1863, General Hooker introduced distinctive insignia for each Corps of the Union Army. They were worn in felt on the top of the kepi hat of infantry soldiers and on the left breast in either cloth or metal by officers. The symbol of the Fifth and Nineteenth Corps was a Cross Pattée. The Army version of this cross had flat sides identical to that of the first "Maltese cross" when it was introduced by the Department in 1865 as a hat badge.

The recruiter of New York firemen into a single regiment was Col. Elmer Ellsworth. It was his pre-war Zouave cadets that drew the attention of the New York firemen and their visit to New York City that drew Ellsworth's attention to the firemen. Soon after the formation of the First New York Fire Zouaves, Ellsworth became the first Union officer to be killed in the war. Sorrow over his death reached far and wide. Lincoln, a close friend of Ellsworth, was deeply affected by his death. Ellsworth's body laid in state both at the White House and at New York City Hall. A new regiment, *Ellsworth's Avengers* (44th New York Volunteers) was quickly organized. The regiment was a part of the Fifth Corps whose symbol was the Cross Pattée, commonly called the Maltese cross.

John Cregier may have played a role in introducing the Maltese cross as the symbol of New York firefighters.

Who introduced the Maltese cross into the Department? Although there are no records that address this, one (or more) of four individuals may have played a role.

John A. Cregier was an Assistant Chief Engineer in the NYFD when he enlisted in the 11th New York Volunteers. He served as the third in command under Col. Ellsworth, with the rank of Major. Cregier's military service was short. He resigned in 1861 and returned to New York and his position in the Department. Cregier held several important posts with the Department such as Inspector of Combustibles and head of the New York Fire Patrol. In 1865, Cregier was appointed by the first Board of Commissioners to serve on a committee to make recommendations about the newly formed paid Department. Is it possible that the recommendation for the Maltese cross badge style, the symbol worn by *Ellsworth's Avengers*, was made by him? There are many obscure facts about Cregier. He is one of the few real-life firemen depicted in the famous Currier & Ives lithograph series, "The Life of a Fireman." He appears in the one entitled, "Facing the Enemy." It is not known if any other prominent members of the Department served in Ellsworth's regiments or had an influence on choosing the Maltese cross.

Some have suggested another individual, General Alexander Shaler, who was connected with the Army, its use of the Maltese cross and the Metropolitan Fire Department. General Shaler's rank came from his Army service. He is credited with bringing a military format to the Department during his tenure on the Board of Commissioners, including the use of military ranks. General Shaler had command in the Sixth, Seventh and Nineteenth Corps. The symbol of the Nineteenth Corps was the Maltese cross. But, there is no clear reason why he might be partial to this symbol since his posting to the Nineteenth Corps lasted less than six months. Also, General Shaler was appointed to the Board of Commissioners by Governor Fenton in 1867, a full two years after the Maltese cross had been adopted by the Department. In fact, he was not mustered out of the military until August 24, 1865, five months after the Act to create the Metropolitan Fire Department was passed and around the same time the actual transition from the volunteer Department began. Therefore it is unlikely, though not impossible, that he played a role in the selection of the Maltese cross.

Finally, perhaps it was Protection Engine Company 5 that initiated the use of the Maltese cross. One badge used by the company was a Maltese cross of an undetermined date. As a volunteer company, it is highly likely that this badge was in use prior to the formation of the paid Department. But how would the badge used by Protection Engine 5 be the foundation for the badge of the paid Department? The answer may lie with members of the company. Among them was Foreman Murray Ditchet, the police Captain who first suggested the issuance of badges to firemen in 1855. Also from Engine 5 came Joshua G. Abbe, one of the first Commissioners of the paid Department, whose own badge is shown in Chapter 8 (which is noticeably not a Maltese cross).

Maltese cross badge used by NYFD Engine 5. This may be the first Maltese cross used by a fire company.

· *The pentagon-shaped badge was adopted by the NYFD in 1884 and is said to resemble the operating nut on fire hydrants used throughout the City.*

In 1884, the pentagon-shaped badge was introduced for identification of civilian employees. It was conveniently designed with a panel inscribed with the employee's position. The pentagon badges were also instituted for company officers. They were replaced with the Maltese cross with eagle in 1937, at which time the pentagon was once again used only for civilian employees. The pentagon was gradually phased out as civilian positions were also issued the sun-ray Maltese cross with eagle.

According to legend, the pentagon-shaped badge was modeled after the five-sided operating nut on the top of fire hydrants used throughout the City. The first hydrant was installed in New York in 1817 and by 1884 at least one model did have such a five-sided nut. If this legend is true, than this piece of FDNY history will soon be as scarce as the pentagon badge itself. In 1997, new caps known as, "The Custodian Hydrant Guard" were placed on top of the operating nut to obscure it and make it more difficult for unauthorized persons to turn hydrants on, a traditional summertime activity of New York City youth.

Although often overlooked, the badge currently issued to officers and civilians in the Department which displays an eagle on top is a *stylized* Maltese cross. Sharper angles at the points of the four arms replace the gentle rounded curves of the fireman's Maltese cross. By removing the eagle and comparing the silhouettes of the two badges the similarity becomes obvious. The current Department specification for these badges refers to them as, *"...maltese cross in shape, having four sun ray arms extending from a circle approximately 1 5/8" in diameter..."*

So, was it Mr. Crieger, General Shaler, Foreman Ditchet or Commissioner Abbe, who suggested the Maltese cross as the shape of the fireman's badge, albeit a cap device, used by the Fire Department? Was its selection based on its use by the U.S. Army rather than by the Knights of St. John? These questions remain unanswered.

In 1870 the Maltese cross was changed to its current design incorporating the letters F.D.N.Y. in lieu of M.F.D. Except for the method by which the badge was attached to the cap, it's design has not changed in over 125 years.

Some time during the late 1800's there was another badge shape commonly utilized, but it is not referred to in any official documents or Department orders. It is a simple, yet attractive shape, based on a circle with flourishes at the top and bottom, the bottom being somewhat more ornate. Badges of this shape bear such titles as Insurance, Ass't. Foreman, Chief of Battalion, and Commissioner.

Silhouette of standard and sunray Maltese crosses used by the FDNY.

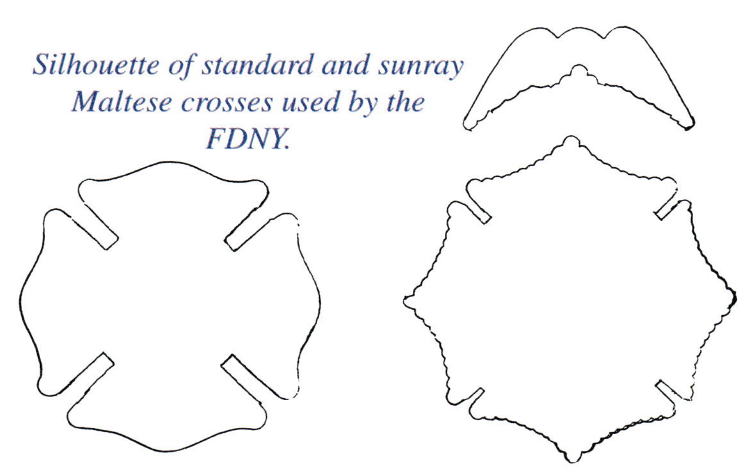

A total of six badges have been issued in an oval shape. Three were for Marine personnel, namely Engineer, Assistant Engineer, and Pilot (see Chapter 7). Three were for special details, namely Theatre Detail, Inspector of Licensed Places of Public Assembly, and Housewatch (see Chapter 9).

Badges and insignia of the FDNY as published in WNYF in 1947.
This is the only published documentation of Department badges.

There has been limited use of other badge shapes, including a unique cross for Engineers and Assistant Engineers of Steamers and a keystone-shaped badge used by the Emergency Auxiliary Corps for only one year during World War I. In 1990, a badge was adopted for the civilian position of Process Server. It is a sunburst-shaped badge similar to badges used by several other NYC agencies, including the Police Department.

In the mid-to-late 1980's Chief of Department, John T. O'Hagen, was presented with a unique badge. It was a sunburst-shape with eagle, similar to that issued by the New York City Police Department. It wasn't until 1996 when this shape was re-introduced for the Commissioner, Deputy and Assistant Commissioners, Chief of Department, Assistant Chief of Department, Chief of Operations, and Chief of Fire Prevention. The intent was to make the badges of the Department's top civilian positions and top uniformed position look similar to those of the analogous positions in the NYPD.

In the early days of the Department, several badges were made from custom dies. These vary widely and were typically designed for high-level positions such as Commissioners. Custom dies have also been used for badges that are not issued by the Department itself but are in some way related to or associated with it.

Badge Numbers

Since the issuance of the first Departmental badges, they have identified in one way or another, the member to whom it is assigned. For the most part, the identification is a distinctive number; letters were used only for Engineers in the pre-1869 period.

With only one exception during the period of 1860 to 1865, the numbers have been large and applied rather than inscribed. In the 1860 set of three company badges, the predominant number was that of the company with the member's number being small and inscribed. In the early days of the Department, all firemen were assigned a number entered into a simple roster maintained by the Common Council. As the Department increased in size, the function of issuing numbers became a bit more difficult. The practice of "retiring" a number eliminated the continued use of many numbers. Now when a new member is appointed, a number is randomly taken from those not currently assigned and the badge with that number is issued.

Retirement of a badge number means that it becomes permanently assigned to the last member who held it and it is never issued again. There is only one reason for retirement of a badge number, when a member dies in the line-of-duty. The first line-of-duty death in the Department was that of Fireman Robert Wintringham of Engine Co. 1 who died on August 24, 1865 when he was run over by a hose tender en route to the famous fire at Barnum's Museum. The practice of retiring badges has not been consistent through the history of the Department. Therefore, some badges of fallen members remain in use.

A member having served thirty years or more may request to keep his or her badge. A written request to retain the badge must be made to the Commissioner (see Appendix). Although members often believe that retaining their badge retires the number, that is not the case. A badge with that number will eventually be made and issued. However, it may take many years before a replacement badge is made and therefore it may be a long time before the number is issued again.

Orders for badges have increased along with the growth in the number of uniformed members. A sufficient number of badges, accounting for retired numbers, has been ordered to accommodate the greatest number of personnel. However, once ordered the highest badge numbers have remained in service. Currently, there are 8,657 firemen in the Department with the highest available badge number of 13486. The following chart illustrates the discrepancy between the current number of uniformed personnel and available badge numbers:

Rank	# of Personnel	Badge # Range
Firefighter	8,657	1-13486*
Lieutenant	1,431	1-3283
Captain	535	1-1239
Battalion Chief	299	1-711
Deputy Chief	49	1-234
Deputy Assistant Chief	2	1-105
Assistant Chief	10	1-113

* Until 1997, firefighters' badge numbers started at 101. In 1997, the badges with the numbers 1 through 100 were issued for the first time.

The badge number 5555 is never assigned. Since the days of telegraph signals in the FDNY, the signal 5-5-5-5 has signified the death of a firefighter in the line of duty. Much like the tradition of not numbering the thirteenth floor of a building, the use of 5555 conjures superstition and is therefore never issued.

Since assignment of badge numbers is random, it is not uncommon or impossible for a member to request a specific badge number. There may be many reasons for wanting a particular number but probably the most common is to have a number that was previously held by a relative. For example, the number 2997 has been held by three generations of the same family who collectively served the Department since 1926, with one family member still on the job today.

Department Name

As described in Chapter 1, the name of the Department has changed several times during its history reverting back to its 1798 name in 1870 - the name by which it is known today. For the most part, badges have been inscribed with the operative name at the time of their issuance or by the appropriate acronym. However, the inscription has not always been verbatim. Since 1870, the acronyms FDNY and NYFD have been used interchangeably.

The first departmentally issued badge in 1855 was inscribed "New York Fire Department." The 1860 changes introduced the acronym "NYFD" on the set of three company-shaped badges. When the Metropolitan Fire Department was created in 1865, the Department name was completely spelled out on the breast badge and abbreviated as "MFD" on the cap device. In 1870, with the conversion back to the 1798 name, Fire Department of the City of New York, the acronym "FDNY" first appeared on badges and other Department items. In 1884, the pentagon civilian badge used an abbreviated form of the official name, "Fire Dept. New York." This abbreviation has been in continual use on civilian and officers' badges through the present.

Recently, another name was introduced, first on the civilian badge for Process Server and later on the other sunburst badges for uniformed members. These currently read, "New York City Fire Department." Although this is the vernacular term used for the fire department and other city departments, it is unofficial and is not used in conjunction with any other official Departmental matters or items.

A subtle difference can be noticed on the enamel ring of sun-ray Maltese cross badges on either side of the Department name. Between the rank or title and the Department name can be found one of three figures, a five-point star, a simple dot, or a small Maltese cross. Of these, the star and the dot are known to be incorporated in dies owned or authorized by the Department. There appears to be no documentation that can pinpoint the dates when these changes were introduced or why. Badges with the small Maltese cross icon are custom-made and hand-cut.

City Seal

The current sun-ray Maltese cross, the recent sunburst badge, and the retired pentagon all have or have had the seal of the City of New York in their center. The City seal appears to have been introduced on Department badges in 1884 with the pentagon. Interestingly, the official City seal has been changed since that time yet Department badges still use the 1884 design.

The current official City seal is described as follows:

> *Arms*: Upon a shield, saltire-wise, the sails of a windmill. Between the sails, in chief a beaver, in base a beaver, and on each flank a flour barrel.

Supporters: *Dexter*: A sailor, his right arm bent, and holding in his right hand a plummet; his left arm bent, his left hand resting on the top of the shield; above his right shoulder, a cross-staff. *Sinister*: An Indian of Manhattan, his right arm bent, his right hand resting on top of the shield, his left hand holding the upper end of a bow, the lower end of which rests on the ground. Shield and supporters rest upon a horizontal laurel branch.

Date: Beneath the horizontal laurel branch the date 1625, being the year of the establishment of New Amsterdam.

Crest: An American eagle with wings displayed, upon a hemisphere.

Legend: Upon a ribbon encircling the lower half of the design the words "Sigilium Civitalis Novi Eboraci" which means, "seal of the City of New York" in Latin.

The whole is encircled by a laurel wreath.

The first seal for the City was developed in 1623 as the seal of the province of New Netherlands. It was modified in 1654 and 1669 with the advent of the city of New Amsterdam and the province of New York, respectively. The seal underwent significant change in 1784 and is the style that is used on FDNY badges. Minor changes occurred in 1915 but were not incorporated into FDNY badges. In 1977, the date on the seal was changed from 1664 to 1625 to recognize the original Dutch settlement rather than the first use of the name New York when the English gained control in 1664.

The seal in the center of Department badges has a rather obvious and unique difference from any of the official seals; the eagle is turned so that one wing is slightly obscured while in the official seal the eagle displays a full forward-facing wing spread. This is the same version of the City seal used on New York Police Department badges and in a variety of applications around the City.

Version of the seal of the City of New York that appears on FDNY badges.

The State of New York seal bears some resemblance to the City seal because it has two people as supporters, a shield as the central device, and the spread-eagle on top of an hemisphere. Beyond these there are many differences. The State seal is not used on any official FDNY badges. In fact, its sole presence signals an unofficial badge. One such unofficial badge displaying the State seal and the lettering, "Canteen - Fire Department New York" was located and is shown in Chapter 18. However, the State seal has been used by at least one department represented in this book, the Stapleton Fire Department of Staten Island. This department was functional prior to the consolidation of the City so displaying the State seal instead of the City seal was entirely appropriate.

Breast Badges and Cap Devices

Badges for the uniformed force are displayed on two areas of the uniform namely, the left breast and the cap. Technically, the cap badge is referred to as a cap "device." (It should also be noted that through the years, the breast badge has also been referred to in Departmental documents as the coat badge and patrol badge.) For non-uniformed positions, only one badge is issued. It would be considered a breast badge if it were to be worn, but it is actually carried in a wallet and used for identification when necessary. At the present time, with the exception of the positions of Fireman and Marine Wiper, all breast and cap badges are distinctly different. For these two positions the difference between their badges is not in design but in structure. The breast badge has a hinged pin and catch mechanism for fastening while the cap device has a threaded post and nut fastener. Early cap badges were fastened using two loops on the back of the badge. The date of change to the current threaded post and nut fastener is not documented.

Cap devices were introduced when uniform caps were introduced, ten years after the first standardized breast badges for the Department were issued. This was in 1865 upon initiation of the paid Department and promulgation of its new rules and regulations. Firemen were issued the "M.F.D." Maltese cross as previously described. However, the first officers' cap devices were cloth. As described for Chief of Department in G.O. #1, April 2, 1879:

> *"Five trumpets measuring one and five sixteenths inches each, crossed with bells outward and projecting beyond mouth pieces so as to form a design one and one-half inches in diameter, to be embroidered in gold upon a circle of dark blue cloth."*

Other Chiefs had a decreasing number of trumpets, which is the case today. For Foremen and Assistant Foremen, the color of the cloth background represented the

Embroidered cap device used by NYFD officers from 1865 through 1890.

type of company the officer was from; specifically, dark blue for engine, scarlet for hook and ladder, and green for chemical. When the first full-time fireboat was introduced in 1875, the background color for marine company officers' cap devices was brown. In addition, company numbers were also embroidered on the badge. The cloth cap devices were converted to metal in 1890 in the FDNY and in 1894 in the Brooklyn Fire Department.

From 1868 through 1907, straw hats were the required head gear around the firehouse during the summer. Specifications for both officers' and members' hats were exacting, even to the size of the braids (5/8" for members and 1/2" for officers). The officer's hat was white while the member's hat was brown.

The original order for these hats required "badges" to be displayed as follows:

> *"Officers will wear the insignia of office now worn upon the fatigue cap sewed upon the front of the hat. The men will wear their Department number in block figures, of white metal, 3/4 inch long, set upon a piece of patent leather 1 3/8 inch wide and 2 3/4 inches long, securely sewed upon the front of the hat, the lower edge touching the brim.*
>
> *The color of the leather upon which the figures are set will be as follows:*
>
> *For men of Engine Companies, Light Blue.*
> *For men of Hook and Ladder Companies, Scarlet."*

These specifications were modified in 1879 to the following:

> *"The registered number in figures of German silver five-eighths of an inch long, fastened on a strip of black patent leather, to be sewed on the front of the band."*

The Department or registered number would be the member's number on the roster of New York firemen.

Trumpets and Axes

The speaking trumpet has long been a symbol of authority carried by the Department's Engineers. The use of a speaking trumpet as an insignia of rank was introduced in the "Bill of Dress" issued by Commissioner Alexander Shaler in January 1868. Reportedly, New York City was the first fire department in the country to use a trumpet as an insignia or rank. The "Bill of Dress" described the use of the trumpets as follows:

> *"...to be embroidered in gold bullion on colored background, dark blue for chief officers, red for hook and ladder company officers, light blue for engine company officers, and light green for chemical company officers."*

In 1890, the axe was chosen as the distinctive symbol for hook and ladder company officers. This coincided with the change to metal cap devices. Since the color background cloth was no longer useful, the axe was a good means of distinguishing officers of hook and ladder companies. By 1890, chemical companies were no longer in service.

Straw hat worn by NYFD firemen from 1858-1907 with number badge.

For company officers, cap devices are oval, silver-colored with one horizontal trumpet or axe for Lieutenant and two vertical trumpets or axes for Captain. Axes are only used for Hook and Ladder company officers while trumpets are used for Engine and Squad company officers, officers without permanent assignment, and all officers assigned to non-firefighting units. Chief officers' cap devices are gold in color, round in shape, and have crossed trumpets in number increasing with rank. Battalion Chiefs have two trumpets, Deputy Chiefs - three, Chiefs-in-Charge, Deputy Assistant Chiefs and Assistant Chiefs - four, and Chief of Department - five.

In 1946, a six-trumpet badge was designed for the Chief of Staff and Operations. However, so many trumpets on so small a badge proved impractical and indistinguishable. Therefore, the five-trumpet design was retained and the four-trumpet design remained to be shared by three ranks, namely Chiefs-in-Charge, Deputy Assistant Chiefs and Assistant Chiefs.

Badge Composition

The metal used for FDNY badges has varied over the years. The earliest badges were made of "German silver," "white metal," and "Prince's metal;" all composite materials comprising a variety of metals.

German silver was used for the Department's first badges. It is composed principally of copper plus nickel and zinc. The proportions of the three metals vary from five percent to forty-five percent for zinc and thirty percent for nickel. At least eighteen percent of nickel is required to give the composition its silver color. During the First World War the name of this metal was changed to "nickel silver," removing any reference to Germany. This metal provides a bright luster to badges struck from it.

White metal is a tin-based (greater than fifty percent) alloy which gives it its silver or grey color. The predominance of tin caused badges to be referred to as "tins." This relatively soft metal was easily die-struck, however it was not very durable. It was also less expensive than nickel silver.

In the 1930's a two-step process was used on Department badges. The base badge was struck in brass and then plated with nickel. This was discontinued during World War II when brass was in high demand for ammunition, but the practice started up again sometime after the war. Current FDNY badges are manufactured from a copper alloy material and are plated with chrome, rhodium, nickel, or gold.

Early badges that were gold in color were manufactured in Prince's metal, an alloy of copper (about eighty-five percent) and zinc (about fifteen percent). It is a variant of brass having a similar but darker hue. Prince's metal is still available today but is known as "low brass" for its

Samples of metals used to make FDNY badges. White metal, Prince's metal, High brass (top to bottom.)

Unfinished FDNY Maltese crosses made from nickel (left), brass (center) and copper (right.)

low zinc content. (High brass, more yellow or gold in color, is comprised of seventy percent copper and thirty percent zinc.) Today, gold-colored badges are struck from the same copper-composite base material as silver-tone badges but are gold-plated to achieve their color.

The significance between the use of the two metals, German silver or white metal and Prince's metal had significance in the Department's first badges. It allowed for immediate distinction between firemen and exempt firemen. In more recent times, gold-color badges are used to represent several specific categories of members and staff. Among the uniformed force, they are issued to Chief officers and to "uniformed" support staff (see Chapter 10). Among the non-uniformed force, they are issued to appointed staff (see Chapter 8) and supervisory positions.

In the 1970's a badge was introduced that had a very unique finish. It was designed for use in the Model Cities Salvage Corps program (see Chapter 17). The badge had a dull finish and appears to look like pewter. However, it is a nickel-plated badge that is treated to give it this "antique" finish. A second badge appeared with this finish in 1990 for Process Servers (see Chapter 10). Both of these badges were manufactured by United Insignia.

Only once in the history of the Department have precious metals been used for badges. These were the badges of Chief Engineer and Engineers in 1857. The badge of Chief Engineer was struck in solid gold and for Engineers in solid silver. Given the weight of these badges (5.25 ounces) at today's prices, it is difficult to differentiate their value based on metal content or rarity. Gold-filling or plating has been used extensively on FDNY badges. Solid gold badges have been issued to several Chiefs of Department but are no longer. Many non-Departmentally ordered presentation badges can be found in solid gold.

Stars and Leaves

The basic Maltese cross with eagle badge remained the same for all ranks and titles, varying only by the title on the ring and the color of the badge, for nearly fifty years. However, in 1971, stars and leaves were added to various badges. This design change was introduced by Caesar Sansevero, Executive Assistant to the Commissioner at that time. Stars and leaves are applied at the foot of the eagle with two clusters of leaves branching out from the center. An appropriate number of stars are added on top of the leaves commensurate with rank or position.

The following is a list of current badges with leaves and stars, along with the number of applied stars:

Rank/Title	Number of Stars
Commissioner	5
1st Deputy Commissioner	4
Deputy Commissioner	3
Assistant Commissioner	2
Chief of Department	5
Chief of Operations	4
Assistant Chief of Department	3
Deputy Assistant Chief	2
Deputy Chief	1
Chief Fire Marshal	2
Deputy Fire Marshal	1
Chief Medical Officer	1

Reportedly, Commissioner Joseph Spinato (1982-1987) ordered a badge with six stars, as he believed that his badge should have more stars than the highest uniformed officer. However, when the Commissioner learned that the Mayor's badge only had five stars, he had his badge changed back to five as well.

Badge Production

Many companies have been contracted to manufacture badges for the FDNY since 1865. Uniformly, a metal stamping process has been used. This requires the development of dies for each badge struck. Dies are a two-part assembly consisting of an embossing die and a force. The embossing die creates the front of the badge while the force creates the rear. The 1855 ordinance stated that, *"said badges shall be struck from dies, which shall be exclusively the property of the Corporation of the Clerk of the Common Council."* Current practice is for the manufacturer to develop and own the dies for all orders. Their die is made from a "hobb" which is a master mold owned and held by the Department. The current pattern of the Maltese cross with eagle includes an enameled ring with the title of the badge holder. Due to the sporadic and relatively infrequent use of some titles, dies with these titles are owned and held by the Department and are loaned to the manufacturer as needed.

Front and back of an official, die-struck FDNY Maltese cross.

Hobb used to create dies for badges.

Die for the enamel ring on a Deputy Assistant Chief's badge.

In order to produce small quantities of badges, for example when a badge has been damaged, "blanks" are maintained by the Department. This eliminates the delay in striking badges from raw material. The badges are simply plated and the numbers applied. Since officer badges also require a die-struck enamel ring, the Department maintains blank badges that have been plated with rings applied for numerous ranks. When needed, the only work to be done is to engrave the member's number. The turnaround time for the process of "finishing" a blank badge is a matter of a few days.

A number of sun-ray Maltese cross with eagle badges appear to be either fakes or errors. The badges were originally manufactured with the rank and Department name stamped directly into the badge in reverse-enamel style. These are found with the ranks of Lieutenant, Captain, and Battalion Chief. This series of badges was manufactured by the C. G. Braxmar Company for the FDNY as part of the first production of the sun-ray Maltese cross badge in 1937. According to the company, because of the large number of badges initially ordered with these titles it would be less expensive to use one-piece construction. For the lesser quantity higher ranks, rings were added over the originals rather than striking completely new blank badges. As a result, when examined from the back, ranks are clearly visible which are not the same as those appearing on the obverse enamel ring. The exact quantity of badges that were made like this and the time period during which they were issued is not certain. Numerous badges from these dies are shown throughout this book.

Badge Distribution

The 1855 Ordinance described the responsibility for assigning and distributing badges, as follows:

> *"The said badges shall be deposited with the Commissioners of the New York Fire Department, who shall be in sole charge of the distribution of the same; and it shall be the duty of the said Commissioners to keep a register of the names of all persons who now are, or who may hereafter become members of the New York Fire Department, and of the number of the company to which said persons are attached."*

In 1868, the task of maintaining and distributing badges was given over to the Property Clerk of the Department. At the present time, the issuance of badges and maintenance of inventory is the responsibility of "The Badge Desk," part of the Bureau of Personnel at headquarters. The "Desk" is staffed by uniformed firefighters detailed to this assignment. Each and every badge produced by the Department is recorded on a card, on which the name of the member(s) to whom the badge is or has been assigned is indicated.

Officer's badges held by the Badge Desk, ready to be assigned a number.

Badges that are broken or otherwise damaged are turned into the Badge Desk for issuance of a replacement. For many years, damaged badges were collected and either shipped to the manufacturer for melt down and reuse of the metal or, in many cases, were dumped into the river. Now they are simply destroyed and discarded.

"Tinning"

The FDNY abounds with tradition. One that seemed to carry much more meaning decades ago was the practice of "tinning" or showing one's badge for certain unique privileges. This may have originated back to the days when firefighters worked twenty-four hours a day with a few short hours off for meals with their families. Although many, if not most, lived within walking distance of the firehouse, others used the NYC transit system to make the short journey. Showing the badge allowed the firefighter to ride the trolley or train free of charge as this was a part of his working day. As the City grew and firemen moved farther away from their assigned stations, they were able to continue commuting to work for free on the City's busses and subways. A 1941 memorandum from the NYC Board of Transportation documents this practice (see Appendix).

A die-stamped rank can be seen clearly on the back of a 1937-vintage Maltese cross with eagle badge.

Another traditional venue for "freebies" were the numerous theaters around New York City. Although there may have been a practical aspect for having firefighters in the audience at theaters, as fires posed hazards in these limited-egress edifices, this was more likely a simple courtesy and "thank you" from the theater owners.

Even with the advent of the two and three platoon systems, safer theaters and commuting by automobile, fireman continued to take advantage of this "perk." Until the fiscal crisis during Mayor Lindsay's administration, members of the Department continued to get free rides on the transit system. Free rides were then limited to only members of the Police Department while in uniform. But many establishments in the City extended courtesies, consisting mainly of discounted prices to their "Bravest."

Although the City, its residents and business owners still admire and respect its firefighters, tinning has become a thing of the past.

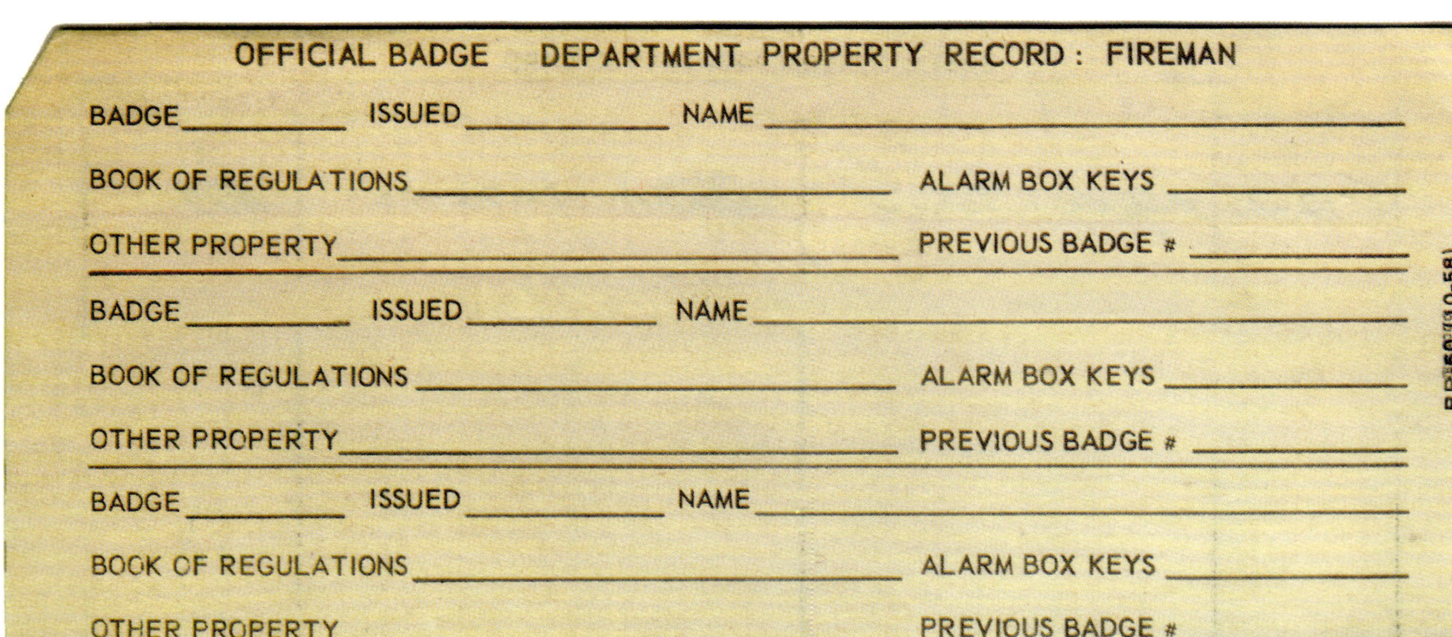

OFFICIAL BADGE DEPARTMENT PROPERTY RECORD : FIREMAN

BADGE ________ ISSUED ____________ NAME ____________________

BOOK OF REGULATIONS ____________________ ALARM BOX KEYS ________

OTHER PROPERTY ____________________ PREVIOUS BADGE # ________

BADGE ________ ISSUED ____________ NAME ____________________

BOOK OF REGULATIONS ____________________ ALARM BOX KEYS ________

OTHER PROPERTY ____________________ PREVIOUS BADGE # ________

BADGE ________ ISSUED ____________ NAME ____________________

BOOK OF REGULATIONS ____________________ ALARM BOX KEYS ________

OTHER PROPERTY ____________________ PREVIOUS BADGE # ________

BP-60 (10-58)

Card used to record information on the issuance of a badge.

CHAPTER 3

New York City Volunteer Fire Departments Past And Present

It was surprising to find so little documentation of the use of badges by New York fire companies. During the volunteer period in New York (Manhattan), there is no record of badges prior to 1855 and very few have been found. In the other boroughs, prior to and after their consolidation into the Greater City, volunteer companies each had distinctive badges. However, records do not indicate the dates of inception. In many cases, badges indicated the company either by name on the badge or by their shape. Badges were not necessarily standardized among companies within the same department.

During the first World War, the Southern New York Volunteer Firemen's Association proposed a plan to the FDNY to activate members of the old volunteer departments in the city's outlying suburbs as a part of the Fire Auxiliary Corps. The Department welcomed the plan and about 800 volunteers were enlisted. Given their experience in firefighting, these volunteers were permitted, in fact requested, to perform actual fire duty unlike other volunteers in the Auxiliary who were restricted from this hazardous task.

At the same time, one full volunteer company was reactivated. In 1918, Bowery Bay Hose Company No. 1 of the North Beach Volunteer Fire Department in Queens applied for a charter that was immediately approved by Commissioner Drennan, Chief Kenlon, Special Deputy Commissioner Joseph, and Mayor Hylan. The area the company served was covered by the paid department, however activation added another company to the Department's war-time initiative.

The geographic area that now constitutes New York City has been served by as many as five hundred volunteer fire companies organized in over ninety departments. As of this writing, there are ten volunteer fire departments still in operation within the city limits. *(See map following page)*

A list of company names, numbers, nicknames, dates of organization and disbandment from the volunteer era are provided in the Appendix to help identify any badges believed to be from the pre-1865 New York companies.

New York (Manhattan)

The first volunteer fire company in New York was organized in 1731. Its Foreman was Peter Rutger. Other companies quickly formed after that. During the following 134 years of the volunteer era, companies would come and go, changing names, nicknames, and official numbers, often reorganizing within days of being disbanded. According to Herbert Asbury author of, *Ye Old Fire Laddies*, volunteer companies often fashioned their own badges.

This is believed to be the oldest badge from a NY fire company- Mutual Hook & Ladder 1 from the early 1800's.

Map of the five boroughs showing communities that had volunteer fire companies.

Mutual Hook & Ladder 1.

A variety of badges from the volunteer era were located and are shown here. The differences among them illustrate the individuality of company designs.

The badge believed to be the earliest one in this book is from Mutual Hook & Ladder Company 1. It is a hand-cut, hand-engraved large shield. Although it is not known when this badge was made and issued, it is thought to be from the early 1800's. Mutual Hook & Ladder was organized on July 10, 1772 and operated throughout the Department's history under that name until absorbed by the paid Department in 1865.

As described in Chapter 2, a very early Maltese cross badge is associated with Engine Company 5. Although the age of the badge cannot be determined, Engine Company 5 was one of the oldest companies in the volunteer New York Fire Department. The badge may represent the first use of a Maltese cross by a fire company.

A beautiful, hand-cut badge from Protection Engine Company 5.

This is believed to be the first Maltese cross used by a fire company, Protection or "Honey Bee" Engine 5.

Early Protection Engine 5 cap device.

Two other badges from Engine Company 5 are shown. The first badge displays the name, "Protection Engine 5." It is presumed to be a very early badge. This presumption is based on the fact that the badge appears to be hand-made. The second badge, which may not be considered a true badge by some collectors, was worn on a hat. The cut-out letters, "Protection - 5" are mounted in an arc on two rails.

Two badges from New York Hose Company 5 are shown. The first is a small badge in the shape of a helmet frontpiece. It displays the initials, "B.S.A." This badge is significant because a helmet with such a frontpiece is shown in Costello's, *Our Fireman*. *Our Firemen* was published in 1887. Therefore, this badge is probably from 1887 or before.

Protection Engine 5 on a run.

Thumbnail badge from New York Hose 5 with the initials "B.S.A." the same as the frontpiece shown in Costello's "Our Firemen" in 1887.

The second badge from New York Hose Company 5 was uncovered in an excavation of a civil war battlefield. It measures 1" high and is made of Prince's metal. As is typical of badges from such excavations, the number is missing but is presumed to have been made of white metal. The banners and finish of the badge bear close resemblance to the NYFD 1855 badge. Hose Company 5 was organized in 1836 under the name "New York" and changed its name in 1860 to "Edwin Forrest."

Badge found at a civil war battlefield excavation, from New York Hose Company 5.

***Thumbnail badges from** New York (Manhattan) volunteer companies; from pre-1850 to 1865.*

A number of small badges, frequently referred to as "thumbnail" badges were made by some companies. They were worn more in the fashion of a lapel pin rather than the uniform display of a badge over the left breast. Many were made of a finer metal than used for typical badges, silver being the most common.

Two badges shown can be traced to their original owners. The first is from Howard Engine Company 34 which was also known as "Red Rover." The badge bears the inscription "Rover - 34" and "J.D. Skidmore."

This badge from Rover Engine 34 bears the owner's name, J. D. Skidmore.

Howard "Rover" Engine 34

C. H. Jenkins and his badge.

An early badge from Hose Company 21, which used the name "Hudson" from 1846 through 1865.

White metal and Prince's metal badges from Americus "Big Six" Engine 6, infamous for one of its early Foremen, William "Boss" Tweed.

The second is a small, simple badge on display at the New York City Fire Museum that belonged to C. H. Jenkins. Measuring a mere 3/4" wide and 1/2" high, it has a simple number on a black background with the company's eagle symbol on top. It is not known which company Fireman Jenkins belonged to, but there were four companies that used the name eagle; Engine 13, Hose 1 and Ladder 4. Fireman Jenkins moved from New York to Chicago where he became a member of Citizen's Fire Brigade, serving as a Lieutenant from 1857 through 1860.

When the paid department was authorized in 1865, there were one hundred twenty-three volunteer companies. Along with the conversion came the assignment of new company numbers and the elimination of their volunteer company names.

Americus ("Big 6") Engine Company 6.

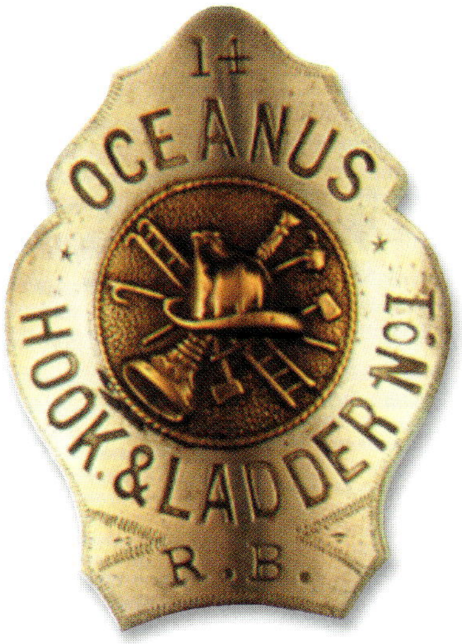

Full-size Manhattan volunteer company badges from Warren Hose 33, Live Oak Engine 44 and Oceanus Engine 11.

Live Oak Engine 44.

Engine Company 38 had two Assistant Foremen. This is the badge for the second Assistant.

Badge from Empire Engine 42 with the title of "Marshal." This is probably from a parade or other special event.

For the volunteer period between 1855 and 1865, a total of four badge styles were issued by the Department to firemen in two distinct periods. These, as well as badges for officers, are described in Chapter 7.

Brooklyn

On December 31, 1768, the Legislature passed an Act, "for the more effectual extinguishment of fires near the Ferry, in the township of Breucklin, in Kings County." However, records indicate that it wasn't until April 7, 1772 that six firemen were actually appointed. The first company to be organized, Washington Engine Company Number 1, occurred in 1785. Firemen were not given exemption from jury and other duties until March 15, 1788 when, like those in New York, they were only exempted during their term of service.

Individual companies served the various areas known as Brooklyn until April 16, 1823 when they were organized into the Fire Department of the Village of Brooklyn. Brooklyn became a city in 1834. Williamsburgh was another city in Kings County and in March of 1823 the Williamsburgh Fire Department was formed. In 1855, the city of Williamsburgh was consolidated into the city of Brooklyn and became known as the Eastern District of the Brooklyn Fire Department. It is important to note that the letter "h" was dropped from the city's name at this time. The Williamsburgh Fire Department comprised ten

Early Brooklyn volunteer fire company thumbnail badges. Pacific Engine 14 of the Village/City of Brooklyn later Brooklyn Fire Department Western District and Excelsior Hose 14 of the City of Williamsburgh later Brooklyn Fire Department Eastern District.

Members of the NYFD Exempt Engine Company gather on the steps of City Hall in August 1893.

Engraving of Atlantic Hose Company 1 of the Brooklyn (volunteer) Fire Department.

EIGHTH ANNUAL BALL
OF THE
FRANKLIN
FIRE ENGINE CO.
No. 3,
MONTAGUE HALL,
BROOKLYN,
MONDAY EVENING, JANUARY 12th. 1852.
GENTLEMEN'S TICKET.

Ticket to the Eighth Annual Ball of Franklin Engine 3 of Brooklyn in 1852.

Although not marked, based on its style this badge is believed to be from Protection Engine 2 of the Brooklyn Fire Department Eastern District.

Bay Ridge Fire Department
(circa 1887-1891)

Flatbush Fire Department
(1821-1896)

Old vollies from Washington Engine of the Flatbush Fire Department on parade in 1930 with their 1852 hand pumper.

Coney Island Fire Department (1886-1895)

New Lots Fire Department (1850-1886)

August F. Herrman of New Lots Fire Department Eagle Ladder 2.

Gerritson Beach Fire Department
(1922-Present)

engine, four hose, and three hook and ladder companies at the time of consolidation. The remainder of the city of Brooklyn formed the larger Western District. Each District had a separate Board of Commissioners, with five in the Eastern District and four in the Western.

As in New York, companies were periodically disbanded and occasionally reorganized. Company names changed over time in some instances. A list of names and numbers of companies of the Eastern and Western Districts are provided in the Appendix. Unfortunately, there is scant information on their dates of organization and disbandment. However, where available, it has been noted.

In 1869, the Brooklyn Fire Department became a paid department and was organized under one Board of Commissioners, having four members. The two Districts were eliminated for all purposes except for their widows' and children's funds. The history of the Department from 1869 is continued in Chapter 4.

Prior to consolidation of Kings County into the Greater City of New York in 1898, other volunteer companies served various communities within the county and what was later to become the Borough of Brooklyn. These included several large departments such as New Lots (1886), Flatbush (1894), New Utrecht (1894), Gravesend (1894), and Flatlands (1896). There were also many smaller departments such as Barren Island, Coney Island, East New York, Plum Beach and Sheepshead Bay.

Flatlands Fire Department
(1859-1897)

Gravesend Fire Department
(?-1895)

During 1895 and 1896, most of these departments became part of the paid Brooklyn Fire Department. In 1898, they were absorbed into the FDNY, although some lingered longer before the areas they served were eventually covered by FDNY companies. Today, one volunteer company continues to serve the Brooklyn community of Gerritson Beach. It operates with one engine company, one rescue company and an ambulance. Earlier, Gerritson Beach F.D. was known as Pioneer F.D.

Badges from the City of Brooklyn Fire Department are shown in Chapter 4. Shown here are badges from other fire departments in what became the borough of Brooklyn as well as some from individual Eastern and Western District companies.

The Gravesend Beach Fire Department circa 1895.

Miscellaneous borough of Brooklyn fire departments.
(l.to r.) Plum Beach (circa 1930), New Utrecht (circa 1896), and Barren Island (circa 1930).

Fireman Pete Young of the Newtown Fire Department.

One of Queens' larger fire departments - Newtown. (?-1913)

Queens

Much like Brooklyn history, the Borough and County Queens was formed over many years until its consolidation into the City. Up until that time, what today is the County of Nassau (which is not a part of the City) was included in Queens. This large area of land was covered by numerous volunteer departments. Upon consolidation of the Greater City in 1898, there were fifteen volunteer fire departments with sixty-three companies, plus the Long Island City Fire Department which was organized as a paid department in 1891. However, by the time the City began to replace the volunteers with paid FDNY units, they had grown to seventy-five companies in seventeen departments.

As communities grew and changed, fire departments also merged or, in some cases were disbanded. For example, at one point the Far Rockaway and Rockaway Beach fire departments were a singular entity known as

Thumbnail badge from Wandowwonnock Engine or Ladder Company of the Newton Fire Department, probably circa 1840s when the company was formed. Other documents related to these companies spell the name "Wandowenock."

Badges from Chiefs of the Jamaica Fire Department.

Valentine Bangert - one of the Chiefs of the Jamaica ***Fire Department.***

Jamaica South Fire Department (1905-1912)

Jamaica Fire Department badges including such interesting ones as "Juvenile" and "Fire Warden."

Maspeth Ladder Company 3 of the Newtown Fire Department in 1908.

the Oceanus Fire Department. The Middle Village Fire Department was absorbed by the Newtown Fire Department in 1889 which did not get taken over by FDNY units until 1913.

Conversion of the majority of volunteers took place from 1898 until 1929. However, several volunteer companies were actually organized in Queens *after* 1898. These included Bayside (1898), Rambersville (1900), Columbia (1901), Douglaston (1903), Rosedale (1904), Black Stump (1905), Little Neck (1905), Creedmore (1907), Little North Beach (1908), Roxbury (1910), Breezy Point (1910), East Hamilton Beach (1910), Jamaica South (1912), Rockaway Point (1916), St. Albans (1916), Forest Hills (1917), and West Hamilton Beach (1928). Five volunteer companies operate in the borough at the present time. They are the West Hamilton Beach Volunteer Fire Department, the Broad Channel Volunteer Fire Department, Rockaway Point Volunteer Fire Department, Point Breeze Volunteer Fire Department, and the Volunteer Fire Department of Roxbury. All are located in the southwest corner of the borough. Several lie on Jamaica Bay adjacent to the JFK International Airport. Rockaway Point and Point Breeze (formerly Breezy Point) lie on the barrier island that forms the bay. Broad Channel is on an island that lies in the bay.

Richmond Hill Fire Department (1890-1913)

Flushing Fire Department (1854-1908.)

Fireman Morris Nelson of the Flushing Fire Department and the cap badge he is wearing in the photograph circa 1905.

Frank Ryan last Chief of the Flushing Fire Department and a close-up of his badge.

Murray Hill Hose Company 4 of the Flushing Fire Dept circa 1905.

Queens Fire Department
(1890-1929)

Whitestone Fire Department
(1871-1929)

Rockaway ***Beach*** *(Oceanus)* ***Fire Department*** *(1882-1905)*

Chief Carl (left) and Assistant Chief Lambert (right) *of the Rockaway Beach Fire Department.*

Rockaway Point Fire Department (1916-Present)

FOURTH ANNUAL BALL
—OF—
Protection Engine Co.
OF ASTORIA, No. 2 L. I. CITY,
At Washington Hall, Monday Evening, Jan. 15, 1872.
DANIEL SULLIVAN, Foreman. JOHN M. FISH, Sec'y.
GEO. WIESENSTINE, Treas. MAURICE LAUGHLIN, Ass't Foreman. Music by Monahan's Band.
Tickets $1, admitting Gent and Ladies

A ticket to the 1872 Annual Ball of Protection Engine of Astoria.

Astoria Fire Department (1839-1873)

Bayside Fire Department (1898-1924)

College Point Fire Department (1857-1908)

Hook & Ladder and Hose companies of the College Point Fire Department circa 1908.

Little Neck Fire Department (1905-1927)

Fire house of the Little Neck Fire Department circa 1905.

Rosedale Fire Department (1904-1927)

Hollis Fire Department (1890-1921)

North Beach Fire Department (1908-1921)

Roxbury Fire Department (1910-Present)

Woodhaven Fire Department
(1887-1929)

Woodhaven Fire Department Engine Company 2 in 1913.

West Hamilton Beach Fire Department
(1928-Present)

Far Rockaway Fire Department (1875-1905)

Elmhurst Fire Department circa 1905.

Corona Fire Department (Pre-1889)

Morris Park Fire Department (Pre-1890)

Broad Channel Fire Department (1908-Present)

Columbia Fire Department (1901-1908)

Staten Island

In its early days, Staten Island consisted of four distinct regions each with its own fire department. Some time around 1850, four areas including South Field, North Field, West Field, and Middletown, consolidated to form the Staten Island Fire Department. Badges from this department copied the style used by the NYFD from 1860 to 1865. Three shapes represented each of the three types of companies in service at that time. This department only endured until the early 1870's when it was dispersed into numerous smaller departments, some of which grew substantially along with the growing population of the island.

At the time of consolidation into the Greater City of New York in 1898, Staten Island enjoyed protection by six volunteer fire departments comprising 56 companies, many of which remained in place until long after the consolidation of Richmond County. Disbandment of volunteer fire companies on Staten Island took much longer than in the other boroughs due to its relative isolation.

North Shore Fire Department
(1844-Present)

North Shore Fire Department (1844-Present) (continued)

Early thumbnail badges from companies in the North Shore Fire Department.

Unidentified Chief of the North Shore Fire Department.

Richmond Engine Co. 1's apparatus circa 1935.
This company is still active today.

Badges from
Port Richmond Engine Co. 3
of the North Shore Fire
Deptartment.

Port Richmond Volunteer Fire Company 3 in 1909.

The island was accessible only by ferry from several neighboring shore points, including Manhattan, until the first bridges were built. The first bridge, the Goethals Bridge from Elizabeth, New Jersey was built in 1928. The first and only bridge from New York, the Verrazano Narrows Bridge from Bay Ridge, Brooklyn was completed in 1964. The Staten Island Ferry is still operated from its original terminal at the southeast tip of Manhattan and the other at St. George, Staten Island. Both terminals have had devastating fires. The fire at St. George terminal in 1946 caused $2,000,000 in damage and thirty-two serious injuries. The Manhattan terminal was the scene of a major fire in 1991. Asbestos installed throughout the building posed a significant health risk to all at the scene and the building was condemned and later rebuilt. The terminal was therefore shut down after the blaze, with ferry operations moving to the neighboring Governor's Island ferry dock, until the entire terminal building was gutted and rebuilt. The New York City Bureau of Marine and Aviation, which operates the Staten Island Ferry, is known to have had its own police department and it is suspected that they may have also had a fire brigade. Two badges that *may* be from this Bureau are shown in Chapter 22.

The disbanding of volunteer fire departments took place gradually from 1904 until 1937. Most were replaced with units from the FDNY by 1905. Today, two volunteer companies continue to serve their communities. Richmond Engine Company No. 1 of Richmond F.D. and Oceania Hook and Ladder Company No. 1 of Travis F.D. operate with one engine company each.

Perhaps the department best represented by badges in this book is the North Shore Fire Department. It was formed in 1874, when the Staten Island Fire Department was disbanded. The North Shore Fire Department comprised the towns of Mariner's Harbor, Castleton, Port Richmond, New Brighton and West New Brighton.

Edgewater Fire Department (1856-1937)

Great Kills Fire Department circa 1908.

The Tompkinsville Fire Police tower of the Edgewater Fire Department was torn down in the early 1900s.

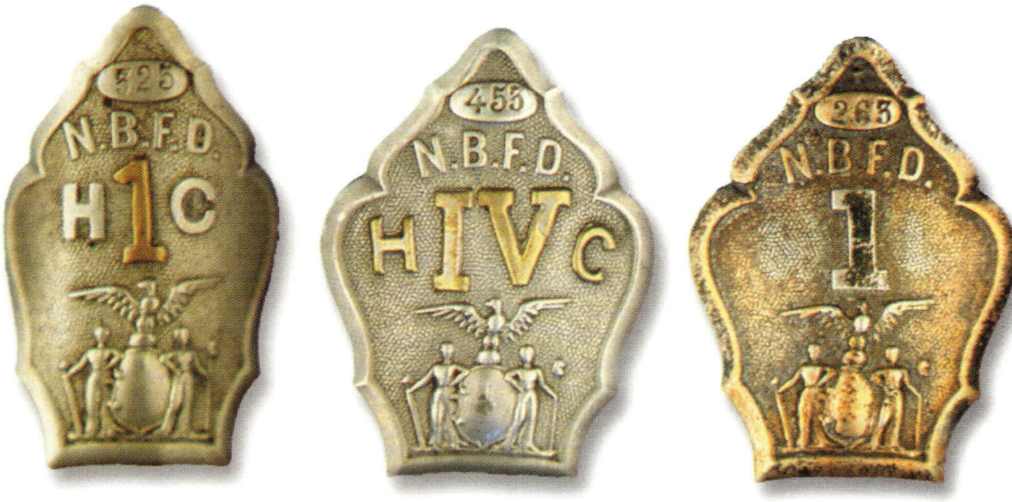

New Brighton Fire Department

Staten Island Fire Department badges in the same style as the 1860-1865 NYFD badges.

Staten Island Fire Department company badges.

An unusual Staten Island Fire Department badge with the title "Alderman" and the Town Hall building in relief.

Stapleton Fire Department
(1856-1875)

*Thumbnail badges **from Protection Hose 3** of the Stapleton Fire Department.*

New Dorp Fire Department
(1860-1915)

Cromwell Engine 8 of Dongan Hills Fire Department
(1903-1912)

Dongan Hills Fire Department circa 1908.

Tottenville H&L Co. 1, circa 1887. (Photo courtesy of J.N. Jacobsen)

Tottenville Fire Department
(1880-1905)

Morrisania Fire Department
(1856-1874)

The Bronx

The borough of The Bronx began assimilation into the City and the FDNY in 1874 when the area south of the Harlem River was annexed. At that time, the Tenth Battalion of the FDNY was formed with six engine companies, two hook and ladder companies plus six pieces of support apparatus (hose wagons and water tenders). The area became known as the 23rd and 24th Wards of New York City. Not much is recorded about the volunteer companies that were disbanded as the result of the takeover. At least one department, the Morrisania Fire Department, had nine companies that were disbanded at that time.

In 1895, New York City extended beyond the Bronx River and into the villages of Wakefield, Unionport, Westchester, Williamsbridge and City Island. These villages had their own volunteer departments, each consisting of one or two companies that were disbanded and replaced by FDNY units. This area of the City was called the 25th Ward.

In the consolidation of 1898, the Greater City of New York took the 23rd, 24th and 25th Wards and additional territory to become the county and borough of the Bronx. Most remaining volunteer departments were replaced by 1899. The volunteers in Eastchester were not disbanded until 1928.

Two companies remains in service in the Bronx, Aviation Hose 3 at Clason Point and Edgewater Volunteer Fire Department in Edgewater Park. Aviation has one engine company while Edgewater operates two mini-pumpers.

Aviation Fire Department
(1922-Present)

Edgewater/Edgewater Camp Fire Department
(1923-Present)

Edgewater Volunteer Hose Company

Aviation Volunteer Fire Company

Eastchester Fire Department
(?-1923)

Tremont Fire Department
(?-1874)

Van Nest Fire Department
(1906-1910)

Wakefield Fire Department
(1890-1896)

Other volunteer fire companies from the borough of the Bronx.

CHAPTER 4

Pre-consolidation City Departments

At two points in the history of the City of New York, paid fire departments from other cities were consolidated into its Fire Department. The first consolidation was in 1865 when the firemen of the City of Brooklyn were to be brought into the newly formed Metropolitan Fire District. However, this marriage was less than amicable. Brooklyn remained a volunteer department and received no benefit from its inclusion in the Metropolitan Fire District. The second consolidation was in 1898 when the borough of Queens was reconfigured and the Long Island City Fire Department was absorbed into the FDNY. It was also in this year that the Brooklyn Fire Department became fully integrated into the FDNY. Both Long Island City and Brooklyn cities were served by volunteer and paid fire departments during their own evolution as described in Chapter 3.

General Order No. 1 of 1898 allowed the insignia, badges, and buttons of the Brooklyn Fire Department to be worn until those of the FDNY could be produced and distributed to the firemen and other members who were incorporated into the FDNY.

The Brooklyn Fire Department's volunteers battling a conflagration in 1848 that leveled a large area of the city.

The Brooklyn Fire Department

Brooklyn was founded as a village in 1816 and became a city in 1834. Earliest records indicate that organized fire fighting efforts began in Brooklyn in 1772, and became more formalized in 1785. Like the New York Fire Department, the Brooklyn Fire Department became a paid service only after the borough had been served for over a century by smaller, volunteer companies.

Legislation to authorize a paid department in Brooklyn was promulgated on May 5, 1869, as "An Act to Reorganize the Fire Department of the City of Brooklyn." This took place four years after the fire departments of the cities of Brooklyn and New York came under the jurisdiction of the Metropolitan Fire District. It resulted from the dissatisfaction of citizens and firemen in Brooklyn who had not received any substantive benefit from their inclusion in the M.F.D. The Act rescinded the jurisdiction of the Metropolitan Fire Commissioners over the Eastern and Western Fire Districts of Brooklyn and at the same time consolidated the independent Boards of Commissioners of those two Districts into one body. It took approximately four months for the conversion from volunteer departments to the singular paid department. The headquarters was established at 365 Jay Street, in a building that stands to this day.

Very early Fireman and Driver badges from the Fire Department of the City of Brooklyn.

Early pinched-shield badges of the Brooklyn Fire Department

Badge registers from the Brooklyn Fire Department dating to 1848 document that the Brooklyn Fire Department issued badges prior to the New York Fire Department and before its consolidation with Williamsburg. Two styles of badges with uncertain dates of use probably come from this period and perhaps before. The first style is a simple, large round badge in brass. Several of these badges were located during research for this book, including one with the title, "Driver" and a company number on it. The second style is a pinched-shield shape. Although the name "Brooklyn" does not appear, it is well accepted by collectors that these "BFD" badges are indeed from Brooklyn. They have been found with only the initials and a number as well as with the word "Insurance."

When the Eastern and Western Districts existed between 1855 and 1869, different badges were issued for each. The Eastern District badge was in the shape of a helmet frontpiece. The Western District badge was in the shape of a frontpiece placed on a round background. Sometime within the same period, the Eastern District used an oval badge circled with a fire hose and topped by a helmet. All of these badges were issued in Prince's metal to firemen and white metal to exempt firemen.

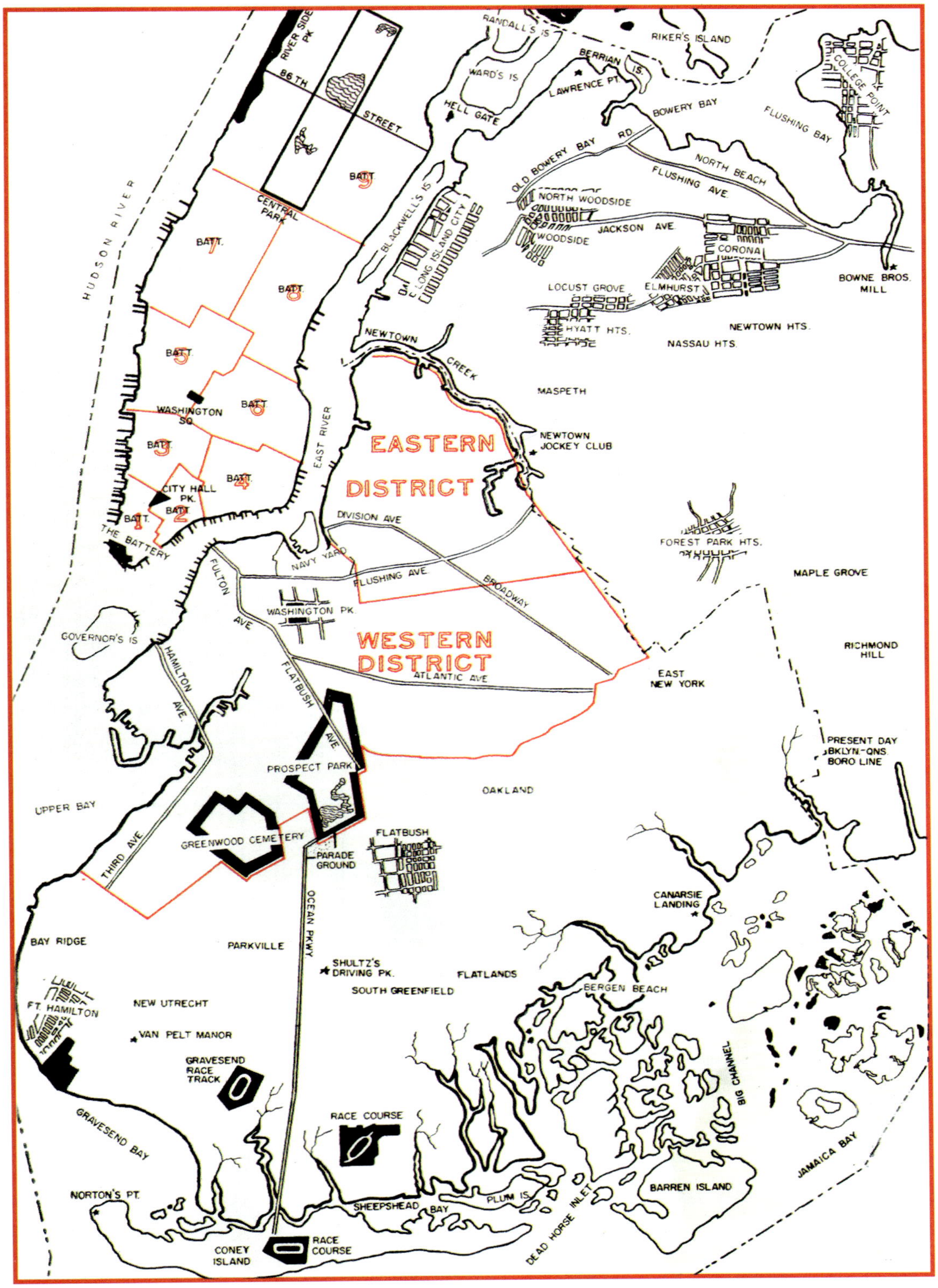

Eastern and Western Districts of the Brooklyn Fire Department as they existed between 1855 and 1869.

Between 1855 and 1869, the Brooklyn Fire Department functioned as two districts, Eastern and Western. These badges identified active and exempt firemen by the metals used.

Badges used by the Brooklyn Fire Department from the 1855 to 1865 have been found in digs for Civil War relics. Firemen of the era that went into battle frequently wore their badge on their Army uniform. The badge of George C. Garmon was found by a fellow solider on his body at the Battle of Williamsburg, Virginia on May 4, 1862. Garmon was a member of the 38th Regiment of New York Volunteers (not a Fire Zouave unit) and Empire Hook and Ladder 3 of the Western District of the Brooklyn Fire Department. The badge was sent to his family and is now on display, along with the letter, at the American Museum of Firefighting.

Badges of Commissioners from the 1855-1869 period.

Two Commissioner's badges from the period prior to 1869 have been located; one each from the Eastern and Western Districts. The oval badge from the Western District, bears the letter "D" and is made of gold. The badge from the Eastern District is round with the words, "Fire Commissioner - Brooklyn Fire Department ED" in raised letters.

Several very interesting and rare badges from the period prior to 1869 are shown. The first is for the Brooklyn Fire Department Board of Appeals. The Board was organized in 1862 to hear any appeals of decisions made by the Board of Commissioners. Little is known about its members but, their badge is illustrated here.

The Board of Appeals was formed in 1862 to review decisions of the Commissioners.

An extremely rare badge for a Trustee of the Brooklyn Fire Department Eastern District circa 1855-1869.

The second is that of a Trustee. There were several Boards of Trustees associated with the Department. There was one main Board of Trustees of the Fire Department which was ultimately responsible for its fiscal matters. Each of the two districts had a Board of Trustees which, after the reorganization of 1869, remained in effect in the Eastern District solely for the purpose of managing the Widows' and Orphans' Fund. The Western District took over the role previously held by the Board of Trustees of the Fire Department. Another Board of Trustees was created in 1866 to set the fire limits, the area that received the service and protection of the Department. The badge shown is from the Board of Trustees of the Eastern District.

Very rare badge for a Fire Warden from the Brooklyn Eastern District.

The third badge is for a Fire Warden of the Eastern District. Although the position of Fire Warden was originally created in February 1836 in what was then the Williamsburgh Fire Department, it had little authority and was often vacant. Several years later, after a rash of vandalism in which hoses were cut at several fires, the position was revitalized and given responsibility for ensuring an adequate water supply for fire fighting and guarding property at fires. The badge illustrated is for a Fire Warden from the Eastern District, the successor to the Williamsburgh Fire Department.

Upon consolidation of the Boards of Commissioners of the two districts in 1869, four Commissioners formed the Board. The number was reduced to three in 1873 and

Brooklyn Fire Commissioner John Ennis and his jewel-encrusted badge.

The most recognized Brooklyn Fire Department badges in "straight B" and "lazy B" styles.

to one in 1880. The fourth person to serve as the single-headed Commissioner of the Brooklyn Fire Department was John Ennis. His badge is described and illustrated in, *Our Firemen: The Official History of the Brooklyn Fire Department* (1892), as follows:

> *"The badge is of heavy Roman gold and is in the form of a circle; in the centre is a monogram of red, white and blue enamel. Above and below the monogram in large raised gold letters, faced in blue enamel, are the words, "John Ennis, Fire Commissioner." A laurel wreath made of Etruscan gold adorns the lower part of the badge, and surmounting it is a solid gold eagle with ruby eyes and a two-carat diamond depending from its beak. Set in the circle at equal distances apart are three more diamonds, each weighing two carats. The reverse side of the badge contains this inscription, "Presented to John Ennis, Esq., Fire Commissioner of Brooklyn, by his friends, February 12, 1886."*

The cost of the badge was reported to be $800, equivalent to approximately $23,000 in 2001 dollars. The whereabouts of the badge are not known.

Fireman Bell of the Brooklyn Fire Department with an equilateral BFD cross on his cap, circa 1898.

Rounded cross or cloverleaf badge.

Standard badge of a Brooklyn Chief.

Restrike from the original die of a Brooklyn Engineer of Steamer cap badge.

Brooklyn Fire Department Headquarters building. Opened in 1892 and, after consolidation with the FDNY on September 23, 1898, continued to be used by the Department until 1984.

Special title cross badges with "Brooklyn" written out. (left to right) Deputy Commissioner, Press, Special and blank.

Badges used after 1869 by the paid Brooklyn Fire Department are believed to be those in the style of a cross. Photographs of the Brooklyn Fire Department just prior to its consolidation into the FDNY in 1898, document the use of the squared, equilateral cross as the last badge to be issued. It is speculated that the rounded, equilateral cross may have been issued to higher ranks or perhaps appointed positions.

Titles in the paid Brooklyn Fire Department paralleled those of the MFD. The uniformed force was led by the Chief Engineer, John Cunningham, who had previously held that post in the Western District. John W. Smith, formerly Chief Engineer of the Eastern District, was appointed Assistant Chief Engineer. The chain of command extended downward through District Engineers (later Battalion Chiefs), to Foremen (Captains) and Assistant Foremen (Lieutenants). Other titles included Superintendent of Horses, Superintendent of Repair Shops, and Medical Officer. Other positions similar to those of the MFD included Engineers and Assistant Engineers of Steamer, Fireman, and Bellringer.

Cap devices and collar insignia for officers of the Brooklyn Fire Department, like those of the MFD, were originally bullion thread on cloth. General Order Number 35, dated October 18, 1894, stated:

> *"On and after this date, the insignia of rank of the officers of the department will be metal, gilded, and shall be worn on the cap and coat collar."*

Two different styles of embroidered cap devices of the BFD circa 1887. James Doyle - District Engineer (left) and Canice Cassin - Inspector (right).

The Brooklyn Fire Department had a Marine Division that began with the commissioning of its first fireboat, the *Seth Low,* in 1886 and the second, the *David A. Boody* in 1893. It is uncertain whether distinctive marine badges were ever issued. No such badges or documentation of them were uncovered.

Full integration of the Brooklyn Fire Department did not occur until the City of Brooklyn's consolidation into the City of New York. In the intervening years, the Brooklyn Fire Department absorbed several volunteer fire departments into its ranks. These departments included the Town of New Lots in 1886, and the Towns of Flatbush, New Utrecht, Gravesend, and Flatlands from December 1895 through February 1896. At the time of consolidation into the FDNY, the Brooklyn Fire Department ranked as the third largest fire department in the United States.

The Long Island City Fire Department

Long Island City is a community in northwestern Queens County; the East River forms its northern and western boundaries, Newtown Creek on the south, and Hazen Street, 49th Street and New Calvary Cemetery on the east. It became an incorporated village in May 1870 and includes the communities of Hunter's Point, Ravenswood, Astoria, Steinway and Sunnyside.

The badges of these volunteer companies were a pinched shield design, not unlike the 1865 MFD badge being used by their cousins in New York City at that time, although significantly smaller in size. They were issued in two varieties, one for firemen and one for exempt firemen. They followed the pattern used in New York City with respect to color and metal composition.

Long Island City Fire Department

This Certifies

That Hurbert Diepold Jr. has been a Member of the Fire Department of Long Island City, and by the faithful and honorable Discharge of his Duties is entitled to all the Exemptions allowed by Law for such services.

In Witness Whereof, We have signed our name hereto and affixed the City seal, this 27th day of October 1891.

Mayor.

Thomas P. Burke City Clerk

Certificate of membership in the Long Island City Fire Department.

Early LICFD company badges.

Early (pre-1891) shield-shape Long Island City Fire Department badges. (l. to r.) engine, hose and ladder company Active Firemen, and Exempt Fireman.

On February 1, 1891, the Long Island City Fire Department was organized as a paid department replacing the volunteers. At that time there were twelve companies operating including eight engine, three hose, and one hook and ladder companies:

Astoria Engine Company
Jackson Engine Company
Protection Engine Company
Franklin Engine Company
Hunter Engine Company
Washington Engine Company
Mohawk Engine Company
Columbia Engine Company
Rambler Hose Company
Steinway Hose Company
Tiger Hose Company
Astoria Hook & Ladder Company

All but one engine and one hook & ladder company remained in operation upon formation of the paid department. By the time of consolidation with the FDNY in 1898, the Long Island City Fire Department had grown to include six engine and three hook & ladder companies.

Long Island City Fire Commissioner (pre-1891) and Assistant Chief (unknown date.)

The badges of the paid department were similar in most respects to the Maltese cross being utilized by the FDNY at that time. However, it retained one feature from the volunteer days.

A piece of apparatus representing the company type is depicted in relief in the lower portion of the central disk. There was also some modification made to the cross. The high points of the arms were made at a sharp angle rather than the rounded curves of the FDNY cross.

The Long Island City Fire Department did not own its own fireboat. It had a contract with the Newtown Creek Towing Company to make use of a boat from its fleet in case of fire on or near the shores of Long Island City. The boat, *Protector,* was outfitted with a fire pump, hose and other firefighting and salvage equipment. No marine badges are known to have been issued.

George Casey, Chief of the Long Island City Volunteer Fire Department.

Long Island City Exempt Fireman from the pre-1891 volunteer era.

Post-1891 Maltese crosses
depicting company type
(l. to r.) Top: hose, ladder,
Bottom: hand-engine, steam engine.

LICFD cap devices.

Engine Company 4 of the Long Island City Fire Department.

CHAPTER 5

Early Associations And Organizations

Beyond the camaraderie shared by firemen within a company, there was a similar camaraderie albeit tainted with a good deal of rivalry, among all members of the fire department. Organizations were formed to facilitate their gathering and, in many cases, act as a unified voice when necessary. The three earliest types of organizations were the Volunteer Firemen's Associations, the Exempt Firemen's Associations, and the Veteran Firemen's Associations, each having their respective qualifications for membership. The sons of firemen also formed fraternal organizations. Later, the need for more formal labor organizations produced such organizations as the Firemen's Mutual Benefit Association and the Fire Department Benevolent Association.

Photographs of firemen from this period often show the men wearing more than one badge. Typically, they were badges from their membership in multiple associations, or their Department badge plus association badge(s).

It was typical practice for firemen to wear badges from both their fire department and their association(s) at the same time.

Volunteer Firemen's Associations

On December 22, 1883, the Volunteer Firemen's Association of New York City was formed. As can be seen by the date, the association was formed after the elimination of the volunteer department. It was created as a means for the "old vollies" to get together and to keep their interest in firefighting alive even after the drastic changes that had occurred in the Department itself.

Assistant Foreman "ECH" of the NY Volunteer Firemen's Association.

The Brooklyn Volunteer Firemen's Association was formed by the consolidation of the Williamsburgh Volunteer Firemen's Association and the Volunteer Firemen's Association of the City of Brooklyn (Western District).

Variations of badges from the New York Volunteer Firemen's Association, organized 1883 and still active in 2001.

Membership certificate of Thomas Moran, Engine Company 9, for the Brooklyn Volunteer Firemen's Association, dated December 17, 1885.

Volunteer associations of local NYC fire departments.

Members in front of the Flatlands Volunteer Firemen's Association building circa 1905.

After consolidation of the City in 1898, the Volunteer Firemen's Association also served as a voice for the volunteer departments throughout the new City that were decreasing in number. A takeover by the FDNY was not necessarily a welcome event in the communities where volunteers served and lived for so many years. The Volunteer Firemen's Association remains in existence to the present day and continues to act as a liaison with the Department for the ten volunteer fire departments that continue to operate within the city limits.

The Volunteer Firemen's Association of the City of Brooklyn - Western District was formed on January 15, 1885. Its name was later changed to the Brooklyn Volunteer Firemen's Association. The Brooklyn Volunteer Firemen's Association appears to be the only organization that issued badges consistent with the color-coding used for departmental badges. These badges were made of Prince's metal or brass for Exempt Firemen and white metal for Firemen.

The archives of the Brooklyn Historical Society contain records of the Flatbush Volunteer Firemen's Association from 1890 through 1939 and the Flatlands Volunteer Firemen's Association from the 1880's through 1925.

Exempt Firemen's Associations

By definition, a volunteer is someone who does not receive compensation for his or her work. So, how do those who benefit from the services of volunteers reward them? Volunteer firemen were made exempt from jury duty and military draft. As early as December 1737, an Act was passed by the General Assembly of the colony stating that members of the Department were "freed, exempted, and privileged from the several offices of Constable and Surveyor of the highways, and of and from being put into or serving upon any juries or inquest, and of and from being compelled to serve in the militia, or any of the independent companies of or in said city, or any or either of them in cases of invasion or other imminent danger." But, this exemption was only for the duration of time they spent on volunteering for fire duty. On April 12, 1816, *"An Act Granting Privileges to Firemen of the City of New York"* granted exemption only after completion of ten years service to the fire department. This service time was reduced to seven years in 1830 and to five years in 1847.

Although exemption from jury duty and military draft is not granted today, the philosophy of rewarding those who have "paid their dues" still exists. Many volunteer fire departments confer exempt status which relieves members from complying with certain departmental requirements like responding to an established quota of alarms.

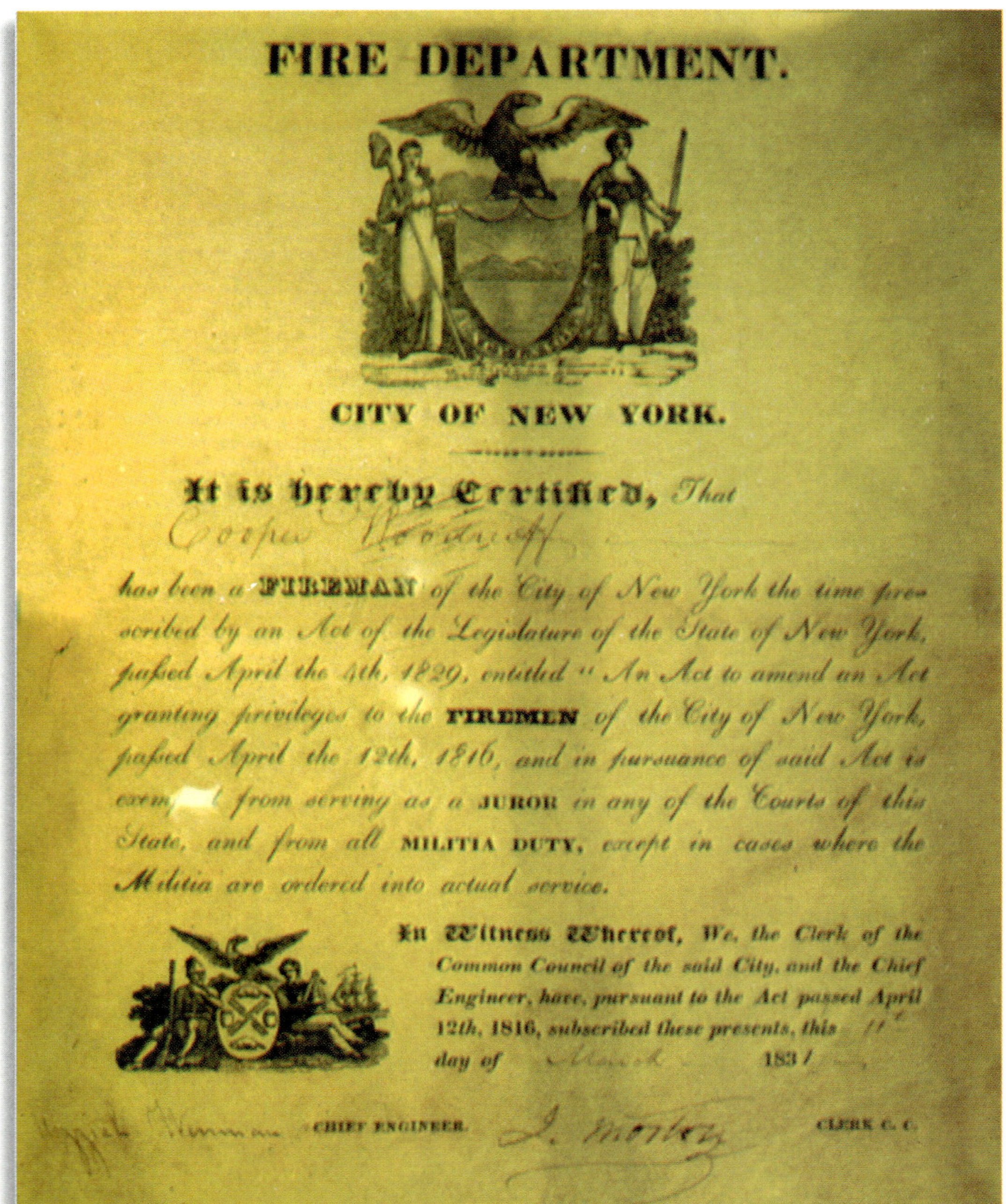

FIRE DEPARTMENT.

CITY OF NEW YORK.

It is hereby Certified, That

has been a FIREMAN of the City of New York the time prescribed by an Act of the Legislature of the State of New York, passed April the 4th, 1829, entitled "An Act to amend an Act granting privileges to the FIREMEN of the City of New York, passed April the 12th, 1816, and in pursuance of said Act is exem[illegible] from serving as a JUROR in any of the Courts of this State, and from all MILITIA DUTY, except in cases where the Militia are ordered into actual service.

In Witness Whereof, We, the Clerk of the Common Council of the said City, and the Chief Engineer, have, pursuant to the Act passed April 12th, 1816, subscribed these presents, this day of 183

CHIEF ENGINEER. CLERK C. C.

Certificate of Exemption for a New York Fireman, circa 1831.

The Association of Exempt Firemen was founded on October 13, 1842. At that time, there was talk of turning the NYFD into a paid organization and its members feared that the Fire Department Fund would be lost to the political powers that were sure to run the Department. An act of the State legislature gave the Exempt Firemen's Association its charter and the assurance that exempt firemen who qualified for financial assistance from the Fund received it. On October 13, 1847 its name was modified to The Association of Exempt Firemen of the City of New York.

Within the New York Fire Department, status as an exempt fireman was noted by a different color badge. In the 1855 and 1860 badge ordinances, the two metals used in badges, namely white metal and Prince's metal, were reversed between the badges of firemen and exempt firemen. A Prince's metal badge with applied numbers of white metal was issued to firemen. The badge for Exempt Firemen was made of white metal with the number in Prince's metal. A similar scheme was adopted by the Brooklyn and Long Island City fire departments.

Once the firemen of the Department became paid, there was no longer a need to provide exemption. As a result, the two reversed color badges in New York, Long Island City and Brooklyn were eliminated.

Association of Exempt Firemen of New York City, founded 1842. Badges believed to be from after 1847.

Exempt Association of the 23rd Ward, was established in 1873 and encompassed the southern half of the Bronx.

Original hand pumper of the NYC Exempt Firemen's Association, which appears on their early badges. This pumper is in the collection of the American Museum of Firefighting at the Fireman's Home in Hudson, New York which was founded by Harry Howard. (Photo courtesy of the American Museum of Firefighting.)

Brooklyn Fire Department Eastern District Exempts were organized in 1882.

Brooklyn Fire Department Exempt badge without number and with a different engine.

Gravesend Exempt Firemen's Association had at least two variation of badges over the years.

Perhaps the most attractive of all the Exempt Association badges is this one from the Westchester Exempts, formed in the midst of a battle to incorporate the northern reaches of the Bronx into New York City.

An interesting pair of badges from the New Lots Exempt Firemen's Association with two different center disks.

Breast and cap from the Town of Newtown Exempt Firemen's Association.

Woodhaven (Queens) and Williamsbridge (Bronx) Exempt Volunteer Firemen's Association buildings.

Woodhaven (Queens) Exempt Volunteer Firemen's Association.

Two companies of exempt firemen (one engine and one hose,) were organized in the NYFD in December 1854 and February 1860, respectively. They were only to respond to greater alarms or other situations calling for additional manpower or equipment.

In Brooklyn, three exempt organizations were formed over time. The first was known as the "Old Guard Association," organized on April 25, 1859 prior to the establishment of the two Brooklyn Fire Department districts. Exempt organizations were subsequently formed in both districts. However, there were significant delays between their initial organization and ultimate incorporation. In the Western District, the Exempt Firemen's Association was organized on January 9, 1852 and incorporated on July 19, 1874. The Eastern District Exempt Firemen's Association was organized on November 14, 1882 and incorporated on April 4, 1883. Between 1899 and 1937, the Gravesend Exempt Volunteer Firemen's Association existed in Brooklyn. The New Lots Exempt Firemen's Association was incorporated in 1886 and existed at least until 1934.

One badge from an exempt organization in the Bronx is shown. It is from the Westchester Exempt Firemen's Association. The association was formed in 1896 by a group of fire companies one year after they were

Cap and breast badges from the Exempt Firemen's Association of Long Island City.

Unidentified member of the Exempt Firemen's Association of Long Island City in 1903.

Souvenir badge from the 1882 firemen's convention in Buffalo given to members of the New York and Brooklyn Exempt Firemen's Associations.

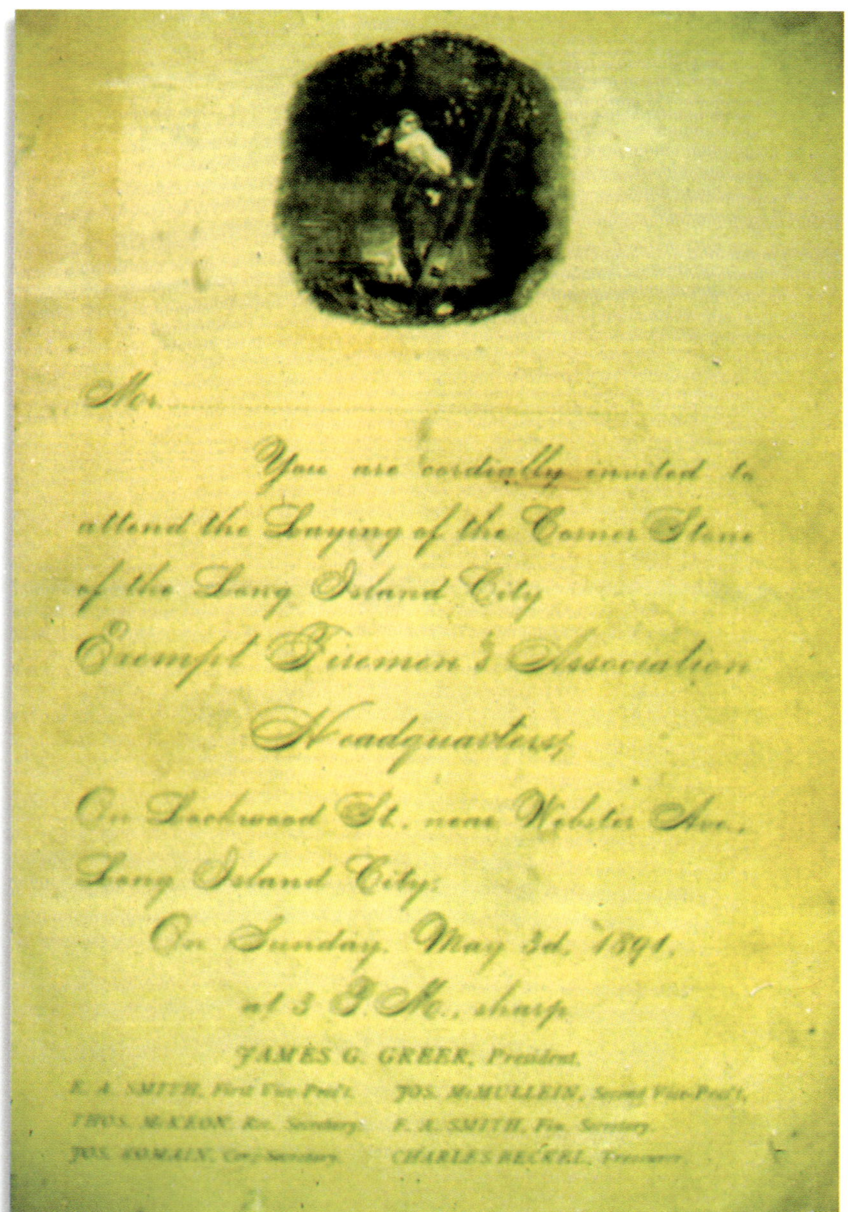

Mr. ....

You are cordially invited to
attend the Laying of the Corner Stone
of the Long Island City
Exempt Firemen's Association
Headquarters,
On Lockwood St. near Webster Ave.,
Long Island City,
On Sunday, May 3d, 1891,
at 3 P.M., sharp.

JAMES G. GREER, President.
E. A. SMITH, First Vice-Pres't. JOS. McMULLEIN, Second Vice-Pres't.
THOS. McKEON, Rec. Secretary. F. A. SMITH, Fin. Secretary.
JOS. ROMAIN, Cor. Secretary. CHARLES BECKEL, Treasurer.

Invitation to the laying of the cornerstone for the LIC Exempt Firemen's Association building on May 3, 1894.

Badges from Exempt organizations of local New York volunteer fire departments.

Aryowius Exempt Firemen's Association of Brooklyn.

disbanded. They were disbanded when the section of the Bronx they served was annexed by New York City. The badge is gold-filled and has brilliant blue enamel. The names of the fire companies that formed the association are displayed on individual ribbons around the circumference of the badge. On top of the badge is a phoenix rising from the flames.

At least one other exempt organization in the Bronx is known to have existed. The Van Nest Exempt Firemen's Association of the City of New York was incorporated on May 18, 1911 one year after the disbandment of the Van Nest Fire Department on May 1, 1910.

The exempt associations died with their members since the only firemen who qualified for membership were the old volunteers. One of the last exempt associations to disband was from the Newtown Fire Department. It was organized soon after the department was formed in 1843. While the department was disbanded in 1913 when the FDNY took over service in the Queens neighborhood, the Newtown Exempt Firemen's Association held its last meeting on April 1, 1970. At that time, its oldest member was 94 years old; the youngest was a mere 78.

It is believed that the Woodhaven Exempt Firemen's Association was the last association to disband, however the exact date is not known. Most, if not all of the other volunteer departments that existed throughout New York had associations for their exempt firemen. Badges for many of them are shown here.

Veteran Firemen's Associations

In addition to Exempt Firemen's organizations, several Veteran Firemen's associations existed in New York and its predecessor cities and towns. The Veteran Firemen's Association of New York City was organized in September 1884. It is still active today. The Veteran Firemen's Association of Brooklyn was organized on

An attractive old badge from the Williamsburgh Veteran Firemen's Association showing the pre-1855 spelling of the city's name.

An interesting, old badge from the Long Island City Veteran Firemen's Association denoting honorary membership.

George Anderson, President of the NYC Volunteer Firemen's Association (1885-1887) and his badge.

The many badges of the NYC Veteran Firemen's Association.

Two versions of badges from the Veteran Volunteer Firemen's Association of the Western District of Brooklyn showing hand and steamer engines respectively.

Badge from the Veteran's Association of the North Shore Fire Department in Staten Island.

November 23, 1886. The Veteran Firemen's Association of the Long Island City Fire Department was organized on March 20, 1890. The veterans had all completed the required term of service in their respective departments. In later years, when a term of service was not specified, a veteran was considered to be any fireman who had completed twenty years of active duty.

A souvenir badge from the firemen's convention in Buffalo in 1891, with the badge of the Veteran Volunteer Firemen's Association of Brooklyn.

Edward Flanagan, President of the Veteran Firemen's Association of Long Island City in 1908.

North Shore (Staten Island) Veteran Firemen's Association building in 1910.

A rare photograph of two members of the Firemen's Sons Association of New Lots, George W. and Louis F. Hurst. The helmet bears the title "Runner Fireman" which is the predecessor to the term "Buff."

Numerous variations of badges from the Volunteer Firemen's Sons Association of New York City.

A Marshal badge from the Volunteer Firemen's Sons Association of New York City.

Firemen's Sons Associations

Another series of organizations that shared a common purpose were the sons of firemen associations. Several of these are known to have been formed in the late 1800's and should not be mistaken for Buff organizations. At that time, Buffs, runners, volunteer organizations - however they were known - were discouraged by the Department and officially banned as of 1841. The "sons" organizations were formed to perpetuate the masculine comradarie and fellowship shared by the fathers. Badges from many of these associations have been found.

Firemen's Sons Association badges from Brooklyn.

The Volunteer, Veteran and Exempt Firemen's Sons Association of New York City was chartered on August 1, 1886. They owned a hose carriage and had uniforms for parades.

Firemen's Mutual Benevolent Association

The Firemen's Mutual Benevolent Association (FMBA) formed in 1893, could be considered the predecessor of today's Uniformed Firefighters Association. It was the Department's first labor union. The mission of the FMBA was to lobby for better salaries and, perhaps more importantly, the two-platoon system. Prior to the formation of the two-platoon system in 1920, each fire company had a set number of members who were relied upon to provide service without relief. Time off was limited to daily meals, infrequent days off, and annual leave. Members resided in the firehouse, only visiting their families during these times.

*Predecessors to the **present-day Uniformed** Firefighter's Association; the Firemen's Mutual Benevolent and Benefit Associations.*

As hard as it would be to accept such working conditions today, the two-platoon system was bitterly opposed by City government. In June 1913, Chief of Department John Kenlon forbade any new member of the Department to join the FMBA. After four years in a struggle with its detractors, the FMBA relinquished its role to the newly formed Uniformed Firemen's Association (See Chapter 15).

Fire Department Benevolent Association

The Fire Department Benevolent Association (FDBA) was formed in July 1900. One of the first actions taken on behalf of its members was to successfully lobby for the repeal of the City charter provisions that would have significantly reduced the pension fund for firemen and instituted the "one day in five" rule. The "one day in five" rule afforded firemen additional time off to spend with their families by giving them one day off out of every five, rather than the established one day off out of every seven. The FDBA was short-lived and its members later formed the core of the Uniformed Firefighters Association (see Chapter 15).

Firemen's Sons Association badges from Long Island City.

Firemen's Sons Association badges from the Bronx.

Firemen's Benevolent Association of Richmond Hill (Queens.)

Fire Engineers Benevolent Association

Not unlike the present day, individual groups within the Department formed to pursue common interests and concerns. The Engineers of Steamers, though active in the Firemen's Mutual Benevolent Association, formed their own benevolent association around 1914. In January 1919, they were chartered as Local 163 of the International Association of Firefighters, a national labor organization that eventually affiliated with the powerful AFL-CIO.

Fire Engineers Benevolent Association insignia.

John J. Perry, President of the Uniformed Engineers Association in 1918.

The logo of the Fire Engineers Benevolent Association was an Engineer of Steamer badge with the letters, "EBA" in place of the word Engineer on the badge, and the letters "FDNY" on the lower panel in place of the badge number. A badge pin used later was based on the traditional FDNY Maltese cross.

Other Organizations

Many of the NYC volunteer companies found it entertaining to form company organizations. These fulfilled a social role beyond that provided by membership in the firefighting company. Some of the organizations had specific purposes, like shooting and hunting sports. Yet, other companies did not form their organizations until after their firefighting days were superceded by the paid department.

There were approximately eighty-two company organizations during the volunteer days of the Department. They had names like The Americus Guards, The Live Oak Volunteers, and the Eagle Cadets. Many companies had more than one club, each with a different purpose. After the paid Department took over, additional clubs were formed by the disbanded companies in order to keep their men together and to give them a social replacement for firehouses. These included the Washington Club, the Warren Association, and the Clinton Engine Company 41 Association. Several badges were found from these early organizations. They are from the Clinton 41, Mazzepa 42 and the Jefferson 26 Associations.

Apparently company clubs were formed in other fire departments around what became New York City. For example, the Nereid Association was formed by Nereid Engine Company 1 of the Wakefield Fire Department in the Bronx on November 24, 1894 and it was incorporated on December 6, 1894. No badges from this or any other company association outside of Manhattan were found.

Two short-lived organizations were the Resigned Firemen's Association and the New York Firemen's Association.

After the Great Fire of 1835, the competence of Chief Engineer Gulick were called to question by the Common Council. They ordered an investigation to be conducted by the Fire and Water Committee which, finally in May 1836, resulted in Gulick's removal. This decision so enraged the firemen of the City that 800 of the 1,500 resigned. They organized themselves into the "Resigned Firemen's Association." William Corp, Foreman of Engine 4 was elected President. The Association also took to political lobbying by backing the Whig party in order to oust the Democrats in control of the Common Council. In the spring of 1837 the Whigs gained control of the Common Council. To fulfill their promise for the firemen's support, they removed Chief Engineer Ryker, who was appointed to replace Gulick by the previous Council, and named Corp as their choice. Corp declined the position

Ball tickets of the Gulick Guards and the Americus Association, social organizations made up of members of NYFD Hose Company 11 and Engine 6 respectively.

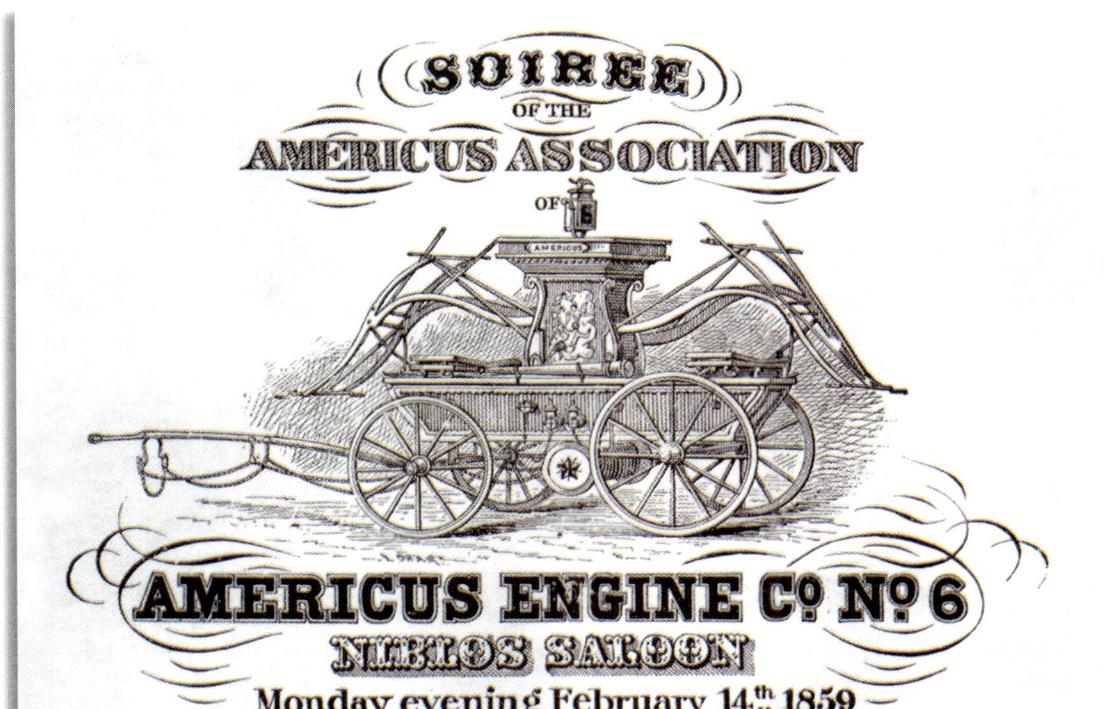

Company associations from the NYC volunteer fire department era.

and Cornelius Anderson was ultimately appointed to the post. This satisfied the resigned firemen and they returned to their companies. The Resigned Firemen's Association had fulfilled its purpose and was disbanded soon thereafter.

Aside for this brief accounting of the Resigned Firemen's Association, little is known of the group. No badges with the Resigned Firemen's Association name have been found, probably because its formation pre-dated the widespread use of organizational badges.

The New York Firemen's Association was formed soon after the institution of the paid Department in 1865 by a number of officers from the old volunteer companies. However, little interest was aroused, membership lagged and the association disbanded in 1876.

CHAPTER 6

Fire Insurance Companies and the New York Fire Patrol

The history of organized firefighting has it roots in fire insurance companies. In England, the companies actually owned and operated the fire brigades. In New York, insurance companies played a different, but equally important role. As early as 1819, the Salamander Society was formed to unify the efforts of the various fire insurance companies in New York. It was the predecessor to the Association of Fire Insurance Companies and to today's New York Board of Fire Underwriters.

Fire insurance companies hold an important place in the fire protection history of New York City. It was not uncommon for fire insurance underwriters to respond to fires. In order to identify the underwriters and to admit them to fire lines, badges were authorized by the Department. Insurance companies also sought to minimize fire damage to insured property by forming the New York Fire Patrol, considered a close cousin of the FDNY.

The following sections describe the badges of fire insurance companies and the New York Fire Patrol.

President of the Williamsburgh Fire Insurance Company with pre-1855 spelling of the city's name.

Fire Insurance Companies

Insurance companies have played a major role in fire prevention and firefighting since the earliest days. In England, fire insurance dates back to the 17th century. In the United States, insurance companies did not establish fire departments but they were instrumental in minimizing the loss of life and property. Given their keen interest in being on hand when one of their insureds was in jeopardy, it was common for insurance officials to respond to the scene of a fire. However, they were not the only civilians to respond. Fires were a common source of entertainment for the general public throughout the eighteenth,

***Badge of the President of the Citizens** Fire Insurance Company.*

Documents from numerous fire insurance companies in New York City during the 18th century.

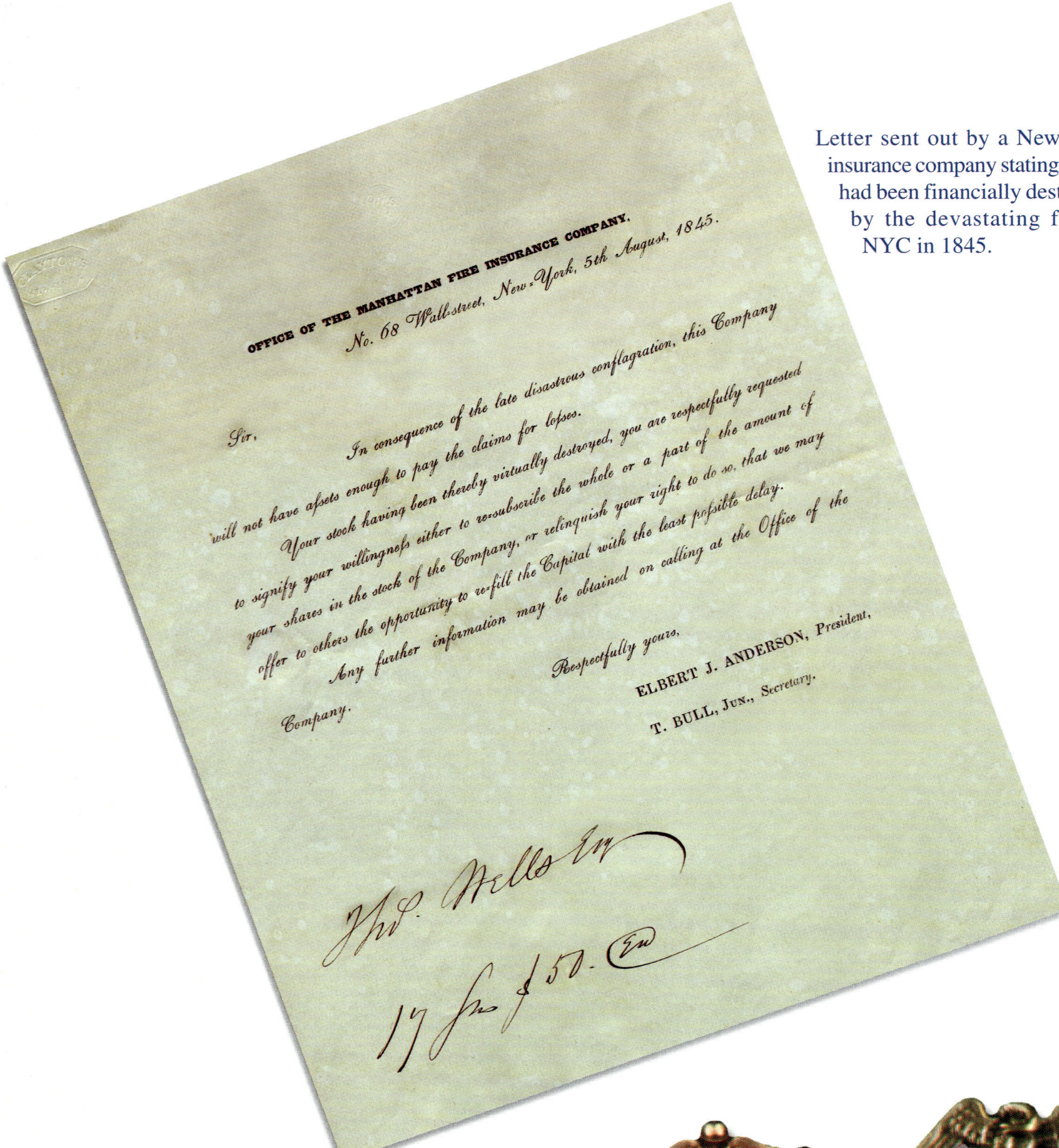

OFFICE OF THE MANHATTAN FIRE INSURANCE COMPANY.

No. 68 Wall-street, New-York, 5th August, 1845.

Sir,

In consequence of the late disastrous conflagration, this Company will not have assets enough to pay the claims for losses.

Your stock having been thereby virtually destroyed, you are respectfully requested to signify your willingness either to re-subscribe the whole or a part of the amount of your shares in the stock of the Company, or relinquish your right to do so, that we may offer to others the opportunity to re-fill the Capital with the least possible delay.

Any further information may be obtained on calling at the Office of the Company.

Respectfully yours,

ELBERT J. ANDERSON, President,

T. BULL, Jun., Secretary.

Thos. Wells Esq

17 shs $50. Ea

Letter sent out by a New York insurance company stating that it had been financially destroyed by the devastating fire in NYC in 1845.

Surveyor - Home Insurance Company - date unknown.

Insurance company "fire lines" badges circa 1870-1884.

nineteenth, and early twentieth centuries. Therefore, it became necessary for civilians with a legitimate purpose for attending fires to get a front row seat. The emerging tradition of issuing and possessing a badge was extended to key members of insurance companies.

In England, insurance company executives were issued coin-like pocket medallions as identification beginning around 1717. In New York, the FDNY began issuing badges to insurance company representatives in 1865. It was policy to issue a badge to the president and secretary of each company in order for them to be admitted to fire lines. The silver badge was round and had the name of the insurance company and the person's title engraved.

In February 1866, the following insurance companies were authorized by the Board of Fire Commissioners to have badges admitting them within fire lines:

Aetna
Arctic
Astor
Atlantic
American
American Exchange
Adriatic
Brevoort
Broadway
Brooklyn
Baltic
Citizens'
City
Columbia
Commercial
Commercial Mutual
Continental
Corn Exchange
Commerce
Croton
Central Park
Eagle
Excelsior
Exchange
Empire City
Firemen's
Firemen's Fund
Fireman's Trust
Fulton
Greenwich
Grocers'
Germania
Globe
Guardian
Harmony
Hamilton
Hanover
Home
Howard
Humboldt
Hoffman
Indemnity
Irving Importers' and Traders'
International
Jefferson
Knickerbocker
Lamar
Lenox
Lorillard
Lafayette
Manhattan
Mechanics'
Mercantile
Merchants'
Metropolitan
Market
Montauk
Morris
National
Nassau
New York Bowery
New York Equitable
New York Fire
New Amsterdam
Niagra
North American
North River
North Western
Pacific
Park
People's
Relief
Republic
Resolute
Royal
Rutgers'
Stuyvesant
Standard
Sterling
St. Nicholas
Star
United States
Union Mutual
Washington
Williamsburg City
Yonkers of New York
Aetna of Hartford
Charter Oak
Sun Mutual

The names of the presidents and secretaries of each of these companies can be found in, *Our Fireman* by Augustine E. Costello, 1887. In addition to the president and secretary, badges were authorized for Fire Insurance Surveyors from Merchants', Brooklyn, Lafayette, and Resolute insurance companies. A later badge for a Surveyor from the Home Insurance Company is shown here.

On May 9, 1867, the New York Legislature granted a charter to the New York Board of Fire Underwriters. After several false starts, as early as June 20, 1850 and again on March 19, 1858, the Board began what became a long history of encouraging fire and loss prevention through proper building standards and safe practices. Much of the Board's work served as a platform for other national initiatives including the National Board of Fire Underwriters and the National Electrical Code. In fact, during its 1858 iteration as the New York Board of Fire Companies, the group argued for and was probably the most influential advocate for replacing the volunteers of the New York Fire Department with paid firemen. The Board took over operation of the New York Fire Patrol in 1867, a role it maintains today. In 1871, it was the Board that loaned the financially strapped City of New York $187,000 to meet the payroll of the FDNY.

A badge was issued by the New York Board of Fire Underwriters sometime prior to 1887. The background of the badge represents rays of light. In the foreground is a tarpaulin, the primary piece of equipment used by the fire salvage corps which represents the protection of property. On the tarpaulin is a cluster of Paxinella leaves. Paxinella is described as a peculiar plant that emits a bright flash of light when struck. On top of the leaves is the monogram of the person to whom the badge is issued. A ring around the edge of the badge has the name, New York Board of Fire Underwriters. The badge is incorporated into the design of the Life Saving Medal issued by the New York Board of Fire Underwriters since 1878.

Badge issued by the New York Board of Fire Underwriters, circa 1895. (Copyright Tiffany & Co. Archives 2002. Not to be published or reproduced without prior permission. No permission for commercial use will be granted except by written license agreement.)

Inspector - New York Board of Fire Underwriters - date unknown.

Until some time in the mid-1900's, the Board's staff conducted fire inspections of insured properties, as well as arson investigations. Staff serving in these roles had badges, some of which are shown here.

One of the New York insurance industry's leaders was Harold V. Smith. Mr. Smith was president of the Home Insurance Company. Aside from his professional involvement with the fire service, Mr. Smith had a keen interest in history and its preservation. He began collecting fire insurance and fire service memorabilia and established a museum in his company headquarters on Maiden Lane. The curator of his museum, Kenneth Holcomb Dunshee, achieved personal recognition as well as recognition for the Home Insurance Company's role in preserving firefighting history. He authored two landmark books, namely *Enjine! Enjine!* in 1939, and *As You Pass By* in 1952. Both publications are known to all fire collectors and served as significant resources in the preparation of this book. The collection of the Home Insurance Company was combined with that of the FDNY resulting in the relocation and revitalization of the NYC Fire Museum in 1984. Items from the collection were also given to the

Badges of Harold V. Smith, President of the Home Insurance Company.

In 1905 the National Board of Fire Underwriters "Committee of Twenty" compiled a report on the status of fire prevention and preparedness in Manhattan and the Bronx.

NEW YORK, N. Y.

REPORT

OF

NATIONAL BOARD OF FIRE UNDERWRITERS

BY ITS

COMMITTEE OF TWENTY

ON THE

CITY OF NEW YORK, N. Y.

MANHATTAN AND THE BRONX.

OFFICERS OF NATIONAL BOARD.

JOHN H. WASHBURN, *President.* GEO. W. BURCHELL, *Vice-President.* CHARLES A. SHAW, *Secretary.*

M. S. DRIGGS, *Treasurer.* H. K. MILLER, *General Agent.*

COMMITTEE OF TWENTY.

HENRY EVANS, *Chairman;* C. F. SHALLCROSS, *Vice-Chairman;*

GEO. B. EDWARDS, E. G. SNOW, GEO. W. BURCHELL, E. C. IRVIN,
FRED W. ARNOLD, HY. W. EATON, R. M. BISSELL, E. L. ELLISON,
C. L. CASE, J. MONTGOMERY HARE, WM. B. CLARK, A. W. DAMON,
ROBT. B. BEATH, E. G. RICHARDS, JAMES NICHOLS, HENRY H. HALL.
WM. N. KREMER, A. H. WRAY, EDWARD MILLIGAN,

Executive Department.

S. H. LOCKETT,

American Museum of Firefighting at the Volunteer Firemen's Home in Hudson, New York. Many of the badges contained in this book are from these collections.

Mr. Smith was recognized for his personal and professional interest by several fire departments, including the FDNY. Three of his personal badges, including his Honorary Deputy Chief's badge (see Chapter 12) and a set of two gold badges issued to him as President of the New York Board of Fire Underwriters from 1940 to 1942, are included here.

Foreman and Report of the Underwriters' Protective Association.

National Board of Fire Underwriters Committee on Fire Prevention, circa 1955.

The "Committee of Twenty" was formed by the NBFU in 1905 and is the predecessor to the Committee on Fire Prevention.

An organization with essentially the same mission and objectives of the New York Board of Fire Underwriters is the National Board. Although its authority and impact is nationwide, not limited to New York City, it is included here for several reasons. First, the National Board's headquarters have always been in New York City, located in the same building as the New York Board. Second, the Board established its "Committee of Twenty" to investigate the status of fire prevention and preparedness in the City of New York. The Report, while not at all scathing, may not have been warmly received by the Fire Department or the New York Board. The present New York Board is a subsidiary of the National Board.

Two badges made personally by NBFU field personnel circa 1960. They found that police officers at fire lines acknowledged this style badge much quicker than the official NBFU-issued badges.

New York Fire Patrol

The history of the New York Fire Patrol parallels that of the FDNY. Its official date of origin is somewhat debatable, but activity is first recorded on March 3, 1762. The "Hand in Hand" fire company was formed on this date. This company and one other, the "Heart to Heart" fire company, did not fight fires like other companies. Rather, they valiantly endeavored to protect personal property from harm. It is not coincidental that the name, "Hand in Hand," was also the name of a fire insurance company in England founded in 1696. The minutes of the Hand-in-Hand Fire Office dated, August 15, 1699, document the first directive to issue badges as the "mark of the office of *waterman*" (e.g., fireman).

Back in the United States in 1803, the Mutual Assistance Bag Company was organized by sixty-five prominent residents of Manhattan. Its purpose was the following:

> *"Upon an alarm of fire every member, with his hat and bags, shall repair to the house or store most in danger belonging to any of the company, and endeavor to the utmost of his power to save their effects by carefully conveying them to some convenient and proper place, and to prevent the embezzlement of the same; and in case no such house or store belonging to any of the company be in danger then each member shall give his assistance where he deems it most proper."*

The practice of protecting and saving property during a fire remained a volunteer effort until 1835 when the Association of Fire Insurance Companies (AFIC) hired four men to "protect [their] interest by preserving property exposed to fire and damage thereto by water."

A dramatic illustration of the New York Fire Patrol springing into action in a Harper's Weekly *article, March 1876.*

The Fire Patrol, as we know it today, became a formal entity in 1839 when the AFIC employed forty men to form the Fire Police Force, twenty-six years prior to the state law establishing the paid fire department. The AFIC was the predecessor of the present New York Board of Fire Underwriters whose Charter in 1867 authorized a corps of men to be known thereafter as the New York Fire Patrol. (Throughout its history, the Fire Patrol has also been known by such names as the Insurance Patrol and Salvage Corps. The latter term was also given to a project of the FDNY funded by the U.S. Department of Housing and Urban Development in 1972. (See Chapter 11.)

Staff of the Fire Patrol consist of a Superintendent and Assistant Superintendent as its chief officers; Captains, Lieutenants, and Sergeants as its uniformed line officers; and first, second, and third class uniformed Patrolmen. Like the FDNY, the Superintendent and Assistant Superintendent are positions appointed, not by the Mayor, but by the Board of Directors of the New York Board of Fire Underwriters (NYBFU) as recommended by its Committee on the Fire Patrol. Promotion to officer ranks is by competitive examination. (A current badge of a member of the NYBFU Committee on the Fire Patrol is shown.)

Badges were first displayed by the Fire Patrol sometime around 1853. Given their rather large size, bowl shape, and the reference to the Patrol's vehicle as a "pie wagon," the badges were given the nickname, "pie plates." This phrase continues to be used today when referring to any oversized badge.

A small, hand-engraved, pinched-shield badge from the "Underwriters Salvage Corps" is displayed in the book, *As You Pass By*, an authoritative source of FDNY history. However, details on the time frame of this badge and its use could not be determined.

Early Underwriters Salvage Corps badge of unknown date appearing in As You Pass By.

In 1882, a regulation uniform was adopted for all Patrolmen and officers including a breast badge in an oval shape depicting a horse-drawn Fire Patrol wagon responding to a fire. For Patrolmen, the white metal breast badge displays their number in the lower portion. For officers, the badge is gilt with a panel in place of the Patrolman's number, and the rank is engraved and enameled. (The rank of Sergeant was introduced on February 20, 1884.) The cap badge of this period was identified as a "double Maltese cross." The significance of this design is not known. The Patrolman's number is applied to the center disk surrounded by the raised words, "FIRE" above, and "PATROL" below.

New York Fire Patrol breast badges
1882-1950.

New York Fire Patrol cap devices
1882-Present.

New York Fire Patrol cap
device, pre-1890.

Fire Patrol badge of the
President of the New York
Board of Fire Underwriters.

New York Fire Patrolman
circa 1910 (Photo courtesy
of Peter Keller.)

New York Fire Patrol officer breast badges
1950-Present.

Superintendent of NYFP *and Secretary of the NY Board of Fire Underwriters - circa 1998.*

NY Board of Fire Underwriters Committee on the NY Fire Patrol - circa 1998.

Quarters of Fire Patrol No. 5

Brooklyn Salvage Corps pie plate circa 1895.

Brooklyn Salvage Corps badges *circa 1895-1910.*

In 1950, the breast badges of the NYFP were changed to match the design of the cap badge, the double Maltese cross. At the same time, the badges for Sergeants, Lieutenants, and Captains were changed to brass while Chiefs' badges remained gold plated. These badges are still in use today.

On February 20, 1884, the office of Medical Examiner for the NYFP was established. Dr. Homer H. Warner was the first appointee. The title of this position was later changed to "NYFP Surgeon." No badges with this title were found but are believed to have been issued.

The Fire Insurance Salvage Corps was organized in the City of Brooklyn in 1895. In 1910, twelve years after consolidation of the fire departments, the Board of Directors of the Fire Insurance Salvage Corps of Brooklyn

Breast badges of the New York Fire Patrol in Brooklyn circa 1916-1930.

Patrolman Louis Haag of the Brooklyn Fire Insurance Salvage Corps, circa 1910. (Photo courtesy of Arthur Smith.)

turned operational control over to the New York Board of Fire Underwriters. The unit numbering used by the Brooklyn Fire Insurance Salvage Corps was a continuation of the New York Fire Patrol's numbering system. Several other changes were made to make the two groups consistent. However, badges for Patrol members in Brooklyn had one subtle difference from those of their counterparts in the rest of the City. The badges were identical in all respects except for the addition of a very small letter "B" at the lowest point on the badge. These were most likely used until the Board of Directors was disbanded in 1930 when the Brooklyn Corps became fully integrated into the New York Fire Patrol.

Close-up of the small 'B' at the base of the 1916-1930 badges.

Cap devices of the Brooklyn Salvage Corps circa 1895-1910.

Brooklyn Salvage Patrol 1 in 1909. (Photo courtesy of Arthur Smith)

CHAPTER 7

The Uniformed Force

The very core of the FDNY, as well as every other fire department, is its line firefighting personnel. Badges can be grouped by three categories of uniformed members in the FDNY, namely, firefighters, officers, and marine personnel. Ranks and titles within these groups have changed over the years and not all positions have had badges.

It should be noted that within the Bureau of Fire Investigation, all Fire Marshals are appointed from the uniformed force. However, once appointed they are no longer responsible to maintain or wear a uniform. As such, they are only issued a breast badge, no cap badge. Therefore, for the purpose of this book, Fire Marshals are not included in the category of uniformed fire personnel. They are included in Chapter 10 - Support Positions.

The first official badge issued by the NYFD to active (left) and exempt (right) Firemen in 1855.

Firefighters

The badge that is acknowledged to be the "first" badge issued to members of the FDNY is the 1855 badge. Even though badges had been used by New York firemen prior to 1855, this was the first uniform Departmental badge to be issued. Two versions of the badge were authorized, one for Firemen and one for Exempt Firemen. The distinction was in the reverse of the metals for the badge and the applied number. For Firemen, the badge was made of Prince's metal with numbers in white metal. For Exempt Firemen, the badge was made of white metal with numbers in Prince's metal. The number was the individual member's number as assigned on the roster of firemen. The badges were breast badges only and, as discussed in Chapter 1, were to be worn by all firemen for admittance within fire lines.

Prior to the adoption of uniforms, members wore their badge on the lapel of their civilian clothes.

Series of badges issued in 1860 reflecting the type of company, company number, active or exempt status, and badge number.

In 1860, due to the high number of lost and stolen badges, the standard badge was changed to a series of three badges. Each badge shape represented the company type, e.g., engine, hose, or hook and ladder. Each badge came in versions for Firemen and Exempt Firemen. There was also a change in numbering introduced at this time. The large applied number, made of metal different than the badge, represented the company number. The individual's number was engraved in a small flat disk on the badge.

When the paid department was formed in 1865, badges changed once again. This change probably represents the most historically significant badge design for the FDNY, as well as for fire services worldwide. It was in 1865 that the Maltese cross was introduced. As discussed in Chapter 2, the exact reason why the Maltese cross was chosen and what, if any, significance it held for those who introduced it, are unknown. The Maltese cross has been the symbol of FDNY firefighters since then and has been adopted, with innumerable variations, by hundreds of other fire services.

In 1865, standardized uniforms were also introduced into the FDNY. Members were so opposed to wearing them that the Commissioners also donned uniforms to show their support for the idea. The fireman's uniform included a cap on which the Maltese cross badge was affixed. Based on extensive review of historical records and photographs, it appears that a breast badge was not issued. The breast badge that was issued during this era (1860-1865) was for fire marshals, inspectors, insurance company officials, and reporters. An example of the badge is shown in Chapter 13.

In 1865 New York State took over control of the NYFD and changed its name to the Metropolitan Fire Department. This is the first Maltese cross badge issued to NYC firefighters as a cap device.

Upon acquisition of the first steam fire engines in 1858, firemen were selected and trained to operate the new equipment. To distinguish them from the Chief Officers who were called Engineers at the time, they were referred to as, "Engineers of Steamer." In 1869, the use of the title Engineer for Chief Officers was discontinued. The position of Assistant Engineer of Steamer was established in 1871. Both of these positions were promotions from the rank of Fireman. A third steamer-related position was that of Stoker. Stokers were civilians and received a pentagon-shaped civilian badge. (See Chapter 10)

Engineer and Assistant Engineer of Steamer badge (circa 1879-1958)

Engineer of Engine Company 27 operating his steamer at a working fire on the Bowery.

The first steamer in the Department was assigned to Engine Co. 2. The first Engineer of Steamer was William Boardman. The last steamer, Engine 49, was taken out of service on May 26, 1933 and the last Engineer, Otto Kuntze, retired in 1958, retiring the badge of Engineer along with him.

Engineer J. Walter Homer.

Information on Engineer of Steamer badges for the period from 1858 through 1872 has not been located. From 1858 to 1869, as previously noted, the title of Engineer was held by Chief Officers. It wasn't until October 1872 that the first "self-propelled" engines were acquired, as described in the Department Order cited below. It is possible that during this time, Engineers and Assistant Engineers were issued the same badges as Firemen. Specifications for the cap badge for Engineers and Assistant Engineers of Steamers, as contained in General Order #1, dated April 2, 1879, read as follows:

> *"...a white metal badge...two and three-quarters inches long and two inches high, with a self-propelling steam fire engine in an oval, one and three-quarter inches high and two and one-eighth inches long in the centre, surrounded with an ornamental border, with the word "Engineer," or "Ass't. Engineer" at the top, and the company number in figures below, to be fastened to the cap with an eye under each end..."*

The specified method of fastening was later modified to a screw-post in the center of the badge.

The Engineer's badge does not bear any FDNY lettering. As such, it is difficult to specifically identify any Engineer's badge as being from the FDNY, or from the time period that the Department issued this badge. Badges with the number applied, engraved, and stamped may be found. Any or all of them may have been used in the FDNY. However, the 1926 FDNY Rules and Regulations do specify that the number is to be the "registered number," not the company number as specified in the 1879 order, and that it should be applied in "raised figures" as opposed to engraved.

The same order specified that the breast badge for Engineers was to be a Maltese cross, identical to that for Firemen, except that the number was to be the company number rather than the member number. Given the similarity to the Fireman's badge, it is difficult to recognize such a badge as an Engineer's badge. It would also mean that the Engineers and Assistant Engineers in a company had breast badges that would be identical in all respects, including the number. This may explain the practice of numbering Firemans' badges beginning with the number 101. Any found with the number below 101 would be for Engineers. Photographs from this era, as well as the Department's Rules and Regulations, document the fact that the unique Engineer's badge was worn on both the cap and breast since the early 1900's.

From 1868 through 1907, summer headwear of firefighters consisted of a straw hat as described in Chapter 2. The member's number was placed on a strip of leather in German silver worn on the hat's ribbon. One example of this hat "badge," of sorts, is illustrated.

Number badge worn on firemen's summer straw hats between 1868 through 1907.

On December 26, 1868, military ranks were instituted for all uniformed members of the Department. This included the ranks of Private and Sergeant for Fireman and Engineers of Steamer, respectively. On March 11, 1871 the rank of Corporal was added for Assistant Engineer of Steamer. These ranks were not displayed on the badge. They were used until 1886. Currently, there are six grades of Firefighter within the FDNY, with Probies being Sixth Grade. Progression to Firefighter First Grade is made based on years of service. There is no distinction in badges for these grades.

The next and most significant change in badges for Firemen occurred in 1870 when the name of the Department changed from the Metropolitan Fire Department back to the original 1758 name, Fire Department of the City of New York. It was then that the Maltese cross was introduced with the letters, "F.D.N.Y." which had not been used on badges before. (Earlier badges utilized the

Members of Engine Company 20 appearing with their summer headgear circa 1905.

Maltese cross adopted in 1870 after return of control of the department to the City and change of name to its present Fire Department of the City of New York. Early version of badge.

Contemporary Firefighter's Maltese cross.

"N.Y.F.D." abbreviation prior to 1865, and "M.F.D." between 1865 and 1870.) Specifications for the FDNY Maltese cross as published in General Order #1, dated April 2, 1879 are as follows:

> *"...a white metal maltese cross two inches square, with the letters "F.D." on the top arm, the letters "N.Y." on the lower arm, a hook and ladder crossed on the right arm, and a hydrant on the left arm, and with the registered number of each member in figures three-eighths of an inch long on a black background, in a circle one inch in diameter in the centre."*

No badges have been located with the numbers on a black background as specified, nor are there any photographs showing such badges. It has been speculated that this specification is a typographical error.

In 1899, Commissioner Scanlon decided that it would be advantageous for retired firemen to have badges. This came as the result of the many retired firemen who left the crowd at the St. Patrick's Day parade to battle a fire at the Windsor Hotel. The badge was considered a fire lines badge. For more information and an example of the Retired Fireman's badge see Chapter 13.

Unidentified turn-of-the-century firefighter showing a hat badge with early style numbers.

Fireman demonstrating the proper wearing of the new collar badges in 1960.

Collar badge or device introduced in 1960.

On July 1, 1960, a Maltese cross was added to the collar of Firemen's uniform shirt. (At the same time, effective August 1, 1960, officers added military-style rank emblems on their collars.) The collar cross measures 3/4 inch in diameter and bears the letters "FD" in the center, "City of" and "New York" in the upper and lower arms, respectively. They have a screwpost and nut for affixing to a hole in the collar. In the 1980's firefighters began to wear their company number (e.g., L156) or unit assignment in chrome-plated numbers/letters on the left collar. The collar badge is worn on the right. Although accepted, the practice of wearing a company number or unit assignment on the collar is not officially part of the Department's Uniform Specifications.

Until February 1883, training to be a fireman in New York City was obtained on-the-job. At that time, the first school for firemen was opened in an old sugar warehouse at the foot of West 158th Street at the North River. The school remained there until 1887 when it was moved to the new fire headquarters built on East 67th Street. It was moved several other times; in 1963 to Ward's Island and in 1975 to Randall's Island where it is currently located and known as the New York Fire Academy.

Probationary Firemen attending classes at the Fire Academy on East 68th Street circa 1936. (Photo courtesy of FDNY/Fire Safety Education Fund)

For an undetermined time period, badges were issued to Probationary Firemen. There is limited documentation on these badges, including a description of them in Departmental Rules & Regulations, circa 1926. They are described as a standard Maltese cross with the words, "Ununiformed Fireman" in the center. While no badges with this wording have been found, several versions of a badge with the words, "Probationary Fireman" have existed. The differences in the three styles of Probationary Fireman badges that exist are attributed solely to changes made by the manufacturers. The first was made from the standard FDNY Maltese cross die which was stamped with a center disk resulting in raised letters. It was manufactured by the N.C. Walter Company. The next had an applied domed disk in the center with engraved letters, manufactured by the Stafford Company. The final version was similar to the second except that blue enamel was added to the engraved lettering. There is no hallmark on this badge to identify the manufacturer. All of these badges had applied numbers.

A book published in 1939 entitled, *Fighting Fire* by Captain Burr Layson mentions Probationary Fireman badges as follows:

"The applicant is appointed a Probationary Un-uniformed Fireman and assigned to a fire house by Commissioner McElligott. At the same time he reports to the School. The first day he reports he is photographed for the records, told what training uniform to buy and given a numbered badge which is the same as the fireman's badge with the exception that it carries the word Probationary. Now he starts his sixty days of intensive training."

Probationary Fireman badges are also mentioned in a 1946 document of the NYC Board of Transportation. The exact date they were discontinued is unknown.

Three variations of Probationary Fireman badges and the embroidered version.

An embroidered badge for Probies was used for many years, roughly during the period from the 1940's through the 1970's. A description of the design and use of the badge is contained in the Department's Uniform Specifications of the time:

> *"Maltese cross in shape, approximately 3 inches square, cotton twill background material, background embroidered in light gray rayon thread, outlined in red rayon thread, letters "F.D." on upper arm, "N.Y." on lower arm, approximately 3/8 inch high, a hook and ladder crossed on right arm and a fire hydrant on left arm, with letters "TS" approximately 7/8 inch high in circle approximately 1 1/2 inches in diameter. To conform in every respect to sample on file at Uniform & Equipment Unit."*

Along with the formation of the School of Instruction, General Order #4, June 7, 1883 established the Life Saving Corps of the Fire Department. Since no formalized instruction existed in the FDNY prior to 1883, the Life Saving Corps was essentially a school for working firefighters as opposed to the new recruits. Its purpose was to introduce the advanced techniques of using the newly introduced scaling ladders and other methods of "life saving." This is mentioned here because of confusion caused by a badge from this period. The badge is that of the New York City Life Saving *Service*. The Life Saving Service has its history tied to the U.S. Life Saving Service, one of the early organizations later consolidated to form the U.S. Coast Guard in 1915.

In 1997, a Cadet program was initiated by the Department. Cadets are people interested in joining the FDNY but are too young to take the examination. These young people are trained to work in a variety of capacities within the Department. As of this writing, they are issued only a Department identification card, not a badge.

The significance of the badge to a firefighter is probably no better described than in a report in the New York newspaper, the *Sunday Mercury*. A story in the *Mercury* edition of May 25, 1862 gives an account of Lieutenant John J. Glass, ex-foreman of Engine Company 16 and member of the Second Fire Zouaves, whose last mortal thought was of his NYFD badge.

A similar story comes from the same battle about Fireman George C. Garmon of the Brooklyn (W.D.) Fire Department, as described in Chapter 4. Though from another century, these stories could just as easily be from the present.

Badges for firefighters have remained the same since 1870. (It is also interesting to note that the basic dress uniform has remained the same since August 1, 1926, when the high color coat was modified.) It is difficult to believe that any changes will ever be made to this established tradition. However, only time will tell.

SUNDAY MERCURY

MAY 25, 1862

AN INCIDENT OF THE FIGHT AT WILLIAMSBURG

During the desperate fight of the Second Fire Zouaves (and in fact the whole of the Excelsior Brigade) at Williamsburg, where they contended against a force three times numerically greater than there own, the heroic Glass (ex-foreman of Engine Co. No. 16) fell mortally wounded. Knowing that if the rebels came back upon the ground, according to their thievish and brutal instincts, they would strip the bodies of the Union soldiers of everything valuable, he placed his badge under a log beside him. His supposition was correct; for, being re-enforced, a portion of the Division under Hooker fell back, and the sneaking villains began to search for plunder among our dead.

In a short time the brigade rallied, and drove the enemy in confusion before them. Poor Glass was discovered still living, but fast journeying to "that bourne whence no traveler returns." As his comrades bent over, and watched his eyes closing in death, he kept pointing to the log. None could understand him at first but finally, on examining the spot they found the badge; and contented to know it was safe, he peacefully breathed his last.

Officers

Throughout the Department's history, ranks for officers have essentially undergone only one major change. From the very first days when the Department became organized in 1737 one individual has lead each company under the title of Foreman. The position of Assistant Foreman was created concurrent with the establishment of company size specifications. In 1762, broader Department officer ranks were established, including one Chief Engineer and a staff of Engineers (previously called Overseers) which changed in size as the Department grew. These ranks at the company and Departmental levels remained unchanged for over one hundred years.

Military rank titles were adopted for all uniformed personnel in 1868. While references to Chief Engineer as Colonel, Assistant Engineer as Lieutenant Colonel, and District Engineer as Major were short-lived, the ranks of Lieutenant and Captain for Assistant Foreman and Foreman are still used today. However, there was an overlap in the use of these latter ranks until 1912, when Foreman and Assistant Foreman were permanently replaced with their current titles. Effective with General Order No. 51 on December 15, 1869 the rank Chief Engineer became Chief of Department, three District Engineers became Chiefs of Brigade and the remaining Engineers became Chiefs of Battalion.

Solid gold badge of Chief Engineer Harry Howard.

Engineers of the NYFD in 1857, all of whom received the newly designed badges of solid silver, with Chief Engineer Harry Howard receiving one of solid gold.

It is interesting to note that military ranks for company officers were used during one other period prior to 1868. In 1776, when the City was placed under martial law, the Department was converted to a military organization with the ranks of Lieutenant and Captain adopted for company officers. Chief Engineer Jacob Stoutenburgh reported directly to General George Washington, himself a volunteer fireman in 1750 in Alexandria, Virginia. The ranks reverted back to Foreman and Assistant Foreman at the end of the Revolutionary War.

Probably the most attractive badges ever issued by the FDNY were those for Engineers in 1857. These badges, specifically authorized by an ordinance of the City Council on March 10, 1857, were struck in solid gold for the Chief Engineer Harry Howard and solid silver for the thirteen Engineers. The detail of the Engineers' badges is beyond compare with any other badge produced in that era or since. They measure 3" in diameter, are 1/8" thick, and weigh approximately 5.25 ounces. Unlike other badges of the era, most were individualized with an assigned letter rather than a number. The letter was engraved on the reverse of the badge. One of the surviving badges has the Engineer's name engraved on the face of the badge.

Three of the four Engineers' badges that are known to still exist originally belonged to William Lamb, Peter N. Cornwell and John Brice. Engineer Lamb held this position in the volunteer Department from 1861 until 1865 when he was hired as Superintendent of the Corporation Yard and later Superintendent of Repairs to Buildings in the paid department. His badge has his name engraved on the front. It is not known if William Lamb was related to

Solid silver Engineer badge of 1857 issued to Peter Cornwell, identified by letter engraved on back.

Engineer William Lamb.

Anthony Lamb, the first Chief Engineer of the FDNY from 1732 to 1736. Engineer Cornwell was elected to the position in 1853. Aside from his role in the volunteer Department, he was an Insurance Surveyor and General Surveyor of the New York Board of Fire Underwriters. His badge has the letter "B" inscribed on the back. John Brice was a member of Mazeppa Hose 42 and served the City as an Alderman from the Twentieth Ward. He was an Engineer from 1856 until the takeover by the State in 1865. His badge has the letter "M" inscribed on the back.

*Badge of **Engineer** William **Lamb** (1862-1865)*

Four badges without inscribed letters were issued to Engineers in 1860, namely William Hackett, Stephen Mitchell, John A. McCosker, and Thomas Roe. The fourth existing badge would most likely have belonged to one of these men.

The 1857 Engineer's badge was ornate, heavy, and so valuable that they were impractical for daily use. Therefore, a "working" badge was also issued to Engineers. It is a large, yet simple, pinched shield with the Engineer's name inscribed along with the words "Engineer - NYFD." Two of these badges are shown. They belonged to Bernard Kenney and Elisha Kingsland.

"Working" badges of Engineer Elisha Kingsland (1854-1865) and Bernard Kenney (1863-1865).

Assistant Engineer Elisha Kingsland.

Elisha Kingsland was the first Chief Engineer of the paid department. Kingsland began his fire service career with Hose 32 and was later a member and Foreman of Engine 26. He was an Engineer for eleven years before being elected Chief Engineer in 1865. He remained in this post until December 14, 1869 when he resigned amid controversy and an investigation by the Board of Commissioners. Little has been written of Bernard Kenney. He was a member of Tompkins Hose 16 for only a short period from 1857 through 1862 before becoming an Engineer in 1863. He held this post until the end of the volunteer Department in 1865, at which time he retired.

In the nineteenth century, documentation of officer badges is notably absent except for Engineers. Badges of a unique design have been located with various company (e.g., Assistant Foreman) and Department (e.g. Chief of Battalion) officer titles. This same style badge is known to have been issued to Commissioners. An illustration of Commissioner Joseph Perley's badge in Costello's, *Our Firemen* dates badges of this design to his tenure in that position from 1873 through 1881.

Officers' badges believed to be from the 1870-1884 period.

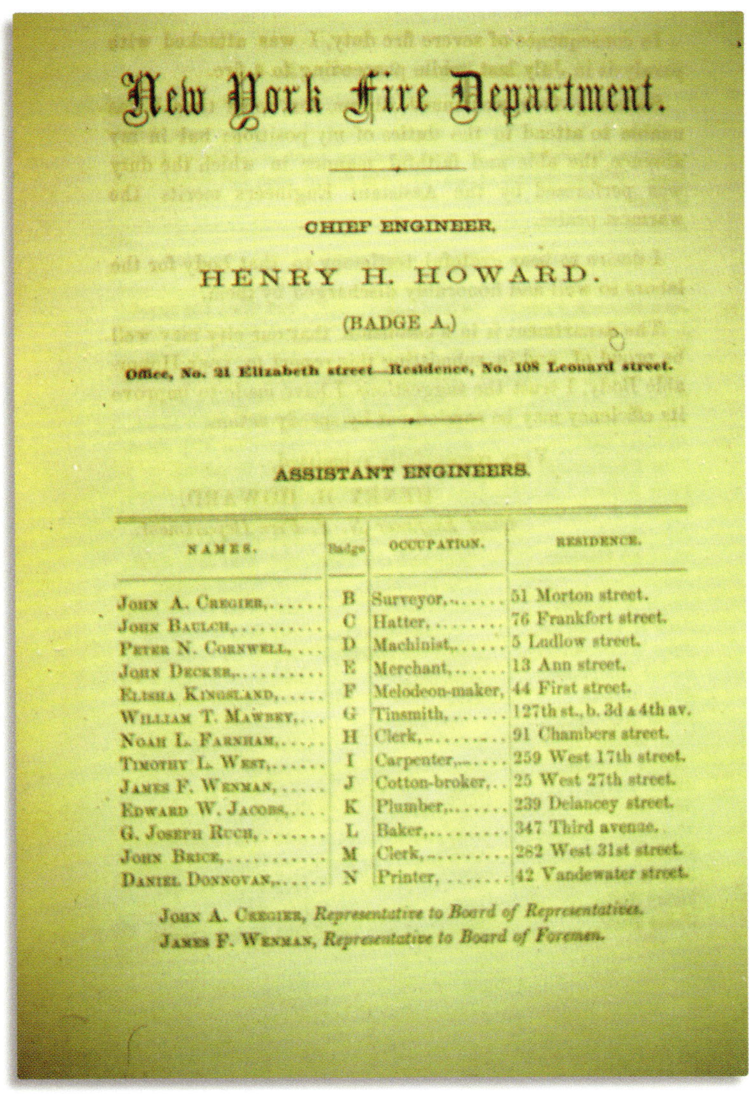

New York Fire Department.

CHIEF ENGINEER.

HENRY H. HOWARD.

(BADGE A.)

Office, No. 21 Elizabeth street—Residence, No. 108 Leonard street.

ASSISTANT ENGINEERS.

NAMES.	Badge	OCCUPATION.	RESIDENCE.
John A. Cregier, ...	B	Surveyor, ...	51 Morton street.
John Baulch, ...	C	Hatter, ...	76 Frankfort street.
Peter N. Cornwell, ...	D	Machinist, ...	5 Ludlow street.
John Decker, ...	E	Merchant, ...	13 Ann street.
Elisha Kingsland, ...	F	Melodeon-maker,	44 First street.
William T. Mawbey, ...	G	Tinsmith, ...	127th st., b. 3d & 4th av.
Noah L. Farnham, ...	H	Clerk, ...	91 Chambers street.
Timothy L. West, ...	I	Carpenter, ...	259 West 17th street.
James F. Wenman, ...	J	Cotton-broker, ..	25 West 27th street.
Edward W. Jacobs, ...	K	Plumber, ...	239 Delancey street.
G. Joseph Ruch, ...	L	Baker, ...	347 Third avenue.
John Brice, ...	M	Clerk, ...	262 West 31st street.
Daniel Donnovan, ...	N	Printer, ...	42 Vandewater street.

John A. Cregier, *Representative to Board of Representatives.*
James F. Wenman, *Representative to Board of Foremen.*

Engineer's roster from 1857 Annual Report indicating badge assignments.

Pentagon-shaped officers' badges (1884-1937)

Chief Edward Croker (1899-1911) in his office circa 1910 and his solid gold pentagon badge.

We know that company officers were issued badges in the pentagon shape for many years. However, they are not mentioned in General Order #1 of 1884 which describes the design and use of this configuration badge. Whether or not use of pentagon-shaped badges for officers occurred before, concurrent with, or after this date is not known.

No Department Orders documenting the changeover to the sunray Maltese cross with eagle badge for officers could be found. However, records of the C. G. Braxmar company confirm that it was introduced in 1937. As described in Chapter 2, this design continues in use today for all company officers and Battalion Chiefs. Department officers' badges are distinguished from company officers by their gold color. For ranks above Battalion Chief, stars and leaves were added and for some selected ranks in 1971. In 1996, a sunburst style badge similar to that used by the New York Police Department was made available for certain Chief ranks.

While current officers' ranks have remained essentially the same since 1869, some above the rank of Deputy Chief have changed over the years. Each change generated different badges which were subsequently retired. For example, the rank of Second Assistant Chief was replaced with that of Deputy Assistant Chief.

One-piece construction Battalion Chief's badge, circa 1937.

Two recent variations of Battalion Chief badges with applied rank/title ring.

Sunray Maltese cross with eagle in chrome-plate for Lieutenant and Captain used from 1937 to present.

Deputy Chief of Department and Deputy Chief-in-Charge.

Variations of Deputy Assistant Chief badges.
Top left - 1937 to 1971, Top right - 1971 to present,
and bottom - Jeweler-made.

Variations of Assistant Chief of Department badges
1937 to present.

Variations of Chief of Operations badges
1971 to present.

Variations of Chief of Fire Prevention badges
1971 to present.

Chief of Training
1971 to present.

Deputy Chief of Staff
circa 1940.

Variations of Chief of Department badges 1937 to present.

Chief of Staff and Operations; a title and badge only issued once, to Frank J. Murphy in 1946.

Chief of Staff and Operations Frank J. Murphy (right) conferring with other officers in 1947.

Miniature officer's badge.

Chief of Department Anthony Fusco (right) wearing a miniature badge. With him are Deputy Assistant Chief Kenneth Cerreta (left) and Commissioner William Feehan (center). The Chiefs are delivering the final fire report on the 1993 terrorist bombing of the World Trade Center. Mr. Feehan (later as First Deputy Commissioner) was killed in the September 11, 2001 attacks that collapsed the World Trade Center towers.

The title of Chief of Staff and Operations was established on December 18, 1946. It was only held by one person in the Department's history, Chief Frank J. Murphy. Within the Department, it is believed that the Chief of Department at that time, Peter Loftus, was a poor administrator and Commissioner Quayle appointed Chief Murphy to function as the *de facto* Chief of Department. On July 30, 1948 all borough commanders were placed directly under his command and their rank was changed from Deputy Chief of Department to Deputy Chief of Staff and Operations, Headquarters Staff. Deputy Chief William Hennessy was designated Assistant Chief of Staff and Operations. A breast badge was issued with the title of Chief of Staff and Operations. Chief Murphy ordered a six-trumpeted cap device to distinguish his position's superiority to the Chief of Department. However, the badge was so cluttered that the number of trumpets was obscured.

In the early 1970's the idea of a miniature badge was introduced for Staff Chiefs. Although these badges were issued, the practice of wearing them was short-lived. Some of the Chiefs liked the miniatures and continued to wear them even though they did not become part of the Department Uniform Specifications. A 1993 photograph of former Chief of Department Anthony Fusco shows him wearing a miniature badge on his uniform. A miniature badge belonging to Commissioner Patrick Walsh (1941 - 1945) was located and is shown in Chapter 8. The use of miniatures during these different time periods suggests that they may have been used periodically throughout the Department's history. The Department no longer issues miniature badges. However, miniature badges may be purchased by firefighters and officers as courtesy badges or lapel pins from a number of vendors (see Chapter 18.)

Cap devices for officers are distinctly different from their breast badge. As described in Chapter 2, cap devices for officers consist primarily of trumpets or axes in a number, configuration, and color that correlates to their rank. For company officers, cap devices are silver with one horizontal trumpet or axe for Lieutenant and two vertical trumpets or axes for Captain. Battalion Chiefs have two trumpets; Deputy Chiefs - three, Chiefs-in-Charge, Deputy Assistant Chiefs and Assistant Chiefs - four; and Chief of Department - five. On all cap devices for Chiefs, the trumpets are crossed. For company officers they are separated.

Embroidered officers' cap devices (circa 1868-1890). Only trumpets were used to show rank so color coding denoted company type; blue for engine, red for ladder.

Cabinet photograph of Assistant Foreman of Engine Company 5 showing an embroidered cap device.

Chief Officer embroidered cap devices (1868-1890).

Generic, FDNY-specification, metal Lieutenant and Captain cap devices without company number (1890-Present).

Chief officer cap devices (1890-Present).

Chiefs' Badges

Chief of Department
Assistant Chief of Department
Second Assistant Chief*
Deputy Assistant Chief
Deputy Chief of Department
Battalion Chief
Chief of Staff and Operations[A]
Assistant Chief of Department in Command[B]
Assistant Chief of Staff*
Deputy Chief of Staff*
Deputy Chief Assistant to Chief of Department*
Deputy Chief of Marine Division[C]
Chief of Operations
Chief of Fire Prevention
Chief of Safety
Chief of Training
Chief in Charge*
Deputy Chief in Charge*
Chief Instructor*
Chief of Support Services**

* - Obsolete
** - Became a civilian position in 1996
A - Only held by one person - Frank J. Murphy - 1946
B - Only held by one person - John J. T. McCarthy - 1940
C - Only held by one person - John Kenlon - 1909-1911

Draeger smoke helmet and Lyle rope gun which appear on cap devices of Rescue company officers. (Draeger Smoke Helmet photo courtesy of FDNY/Fire Safety Education Fund)

Lieutenant and Captain cap devices from Rescue companies displaying the Lyle Rope Gun and Draeger Smoke Mask.

Cap devices of Rescue company officers are silver and display a montage of rescue equipment including a Draeger breathing apparatus, a spool of rope and a Lyle rope gun. The number of rope-guns is indicative of rank, namely one (horizontal) for Lieutenant and two (vertical) for Captain. This unique special unit badge was introduced in 1915.

It has been standard practice to display the company number on Lieutenants' and Captains' cap devices. On a Lieutenant's badge, the company number is affixed above the trumpet or axe. On a Captain's badge, the company number is affixed between the trumpets or axes. Officers of Squad Companies are issued a cap device with trumpets. However, when the company number is applied, it is preceded by the letter "S." These badges are ordered by the company, not the Department. They are kept by the company and passed from officer to officer. In recent years, many companies have gotten away from this tradition.

Lieutenants' cap devices with various company and assignment designations.

Captains' cap devices with various company and assignment designations.

Company officer cap devices can also be found with an assortment of unit designations that may take some deciphering. They include "SP" for Superpumper, "WT" for Water Tower, "HQS" for Headquarters, and "SSS" for Special Services Squad, among others.

A cap device for "Acting" officers is no longer issued. Acting officers are administratively appointed ranks, meaning that the individual is not necessarily qualified for the rank, (e.g., on the promotion list.) Rather, this rank is awarded for prestige and recognition based on the member's assignment. This is particularly true when a firefighter is given an officer designation at the rank of Acting Lieutenant. But, there have been other Acting appointments including Chief ranks. There have been very few such positions in the Department. At the present time there are less than ten.

Acting Lieutenant cap devices no longer in use.

The Acting Lieutenant cap badge, displaying an axe or trumpet, has been documented as far back as 1915. Unlike the company number on officer badges, the word "ACTING" is an integral part of the die and is stamped directly into the badge. At the present time, individuals with Acting ranks are permitted to wear the uniform and insignia of that rank and they are issued a badge as if they were fully vested at the position. (They are not, however, given the pay commensurate with the rank.)

Marine Division

As described in Chapter 1, New York City is surrounded by water. The need to fight fires on ships, wharves and docks was recognized early in the Department's history. The first "floating engine company," as it was called, was formed in 1800. It was a simple flat scow, or barge, with a standard hand-pump engine mounted on-board. The company remained in service until 1814. The boat was manned by a company of volunteer firefighters and, as a barge, did not have a nautical crew. The firefighters did double duty by rowing the boat to the scene.

In 1866, the Department contracted with John Baxter & Sons for the use of their steam salvage tug, *John Fuller*. The tug, which was equipped with steam fire pumps manufactured by Amoskeag, was on-call and would respond when summoned by officer in command at a fire.

On May 12, 1875 the fireboat, *Havemeyer* was placed in service and was staffed by a company consisting of two officers, two engineers, five firemen and a pilot. The fireboat was designated Engine Company 43. It had no firehouse, the crew lived on-board. Most historian consider the *Havemeyer* to be the first official fireboat of the FDNY.

There are four uniformed positions in the Department that are unique to the Marine Division. They are Pilot, Assistant Marine Engineer, Marine Engineer, and Wiper. Initially, they were all civilian positions. The first three were brought into the uniformed force in 1907. The titles have remained the same since the *Havemeyer* was inaugurated in 1875. The Wiper became a uniformed position in 1963 when the title was changed from Marine Stoker.

The Pilot is the captain of the ship for all nautical responsibilities. The Marine Engineer and Assistant Marine Engineer are responsible for the operation of the engines and engine room equipment. The original Marine Stokers, like those assigned to land-based steam fire engines, were responsible for keeping the boiler fueled and operating. The Marine Wiper's responsibilities are for maintaining the operability of all below-deck equipment. The name derives from the task of wiping down the soot and oil films on exposed surfaces in the engine room caused by early coal and diesel engines.

New York's first fire boat, pre-dating the formation of the Marine Bureau.

Marine Engineer and Pilot breast **(pentagon)** ***badges and cap devices (1891-1952.)***

Fireboat Firefighter *as pictured on marine cap devices (1937-1952).*
(Photo courtesy of John A. Calderone)

Badges for Pilot and Marine Engineers also have some current and historic uniqueness. Before each of the positions were brought into the uniformed force, all four were issued a standard, civilian, pentagon-shaped breast badge. Three of these positions are also unique in that they are the only non-officer, uniformed positions whose breast and cap badges differ. Until 1952, the cap badge was a large ornate oval with a fireboat in relief in the center. In the upper curve of the oval was the title. At the widest point of the oval on either side were the letters "F" and "D." At the lower curve were the "N" and "Y" on either side of a panel in which the member's number was engraved. The outer border of the entire badge was finished with a fancy scroll-type edge with flourishes at the upper, lower, and side points.

The Pilot and Marine Engineers badges were manufactured in two distinct variations with regard to the fireboat in the center. The fireboats were either the *New Yorker* or the *Firefighter*. The *New Yorker* was the FDNY's third fireboat. It was commissioned as Engine Company 57 on February 1, 1891 and was replaced by the *John Purroy*

Cap device with relief of the fireboat New Yorker *from original die in the archives of C. G. Braxmar Company.*

Fireboat New Yorker *as pictured on Marine Bureau badges prior to 1937.*
(Photo courtesy of John A. Calderone)

Marine Engineer ***and Pilot breast badges*** *and cap devices (1952-Present).*

Mitchel on January 1, 1917. The *Firefighter* was placed into service on November 16, 1938 as Engine Company 57. It is still in service today as Marine Company 9. The two badges look very different. The earlier *New Yorker* version shows the boat from the bow, at somewhat of an angle. The *Firefighter* version is a side view of the port side of the vessel. The *New Yorker* has a taller stack that operated by natural draft, while the *Firefighter* has a shorter, induced draft stack. The exact date that the Pilot and Marine Engineers' badges were changed is not certain. However, records of the C.G. Braxmar Company, which made the first marine company hat badge, indicate that it was introduced in 1891. It is clear that the die with the *Firefighter* was created sometime after 1938.

The current breast badges for Pilots and Marine Engineers are the sun-ray style Maltese cross with eagle. These are the only non-officer, uniformed positions to whom this style badge is issued. In 1952, the cap badges were changed to an appearance similar to that of company officer-type cap badges. They are still in use today. The raised letters, "F.D.N.Y." are applied at the upper curve and the member's number is engraved on a plate in the lower curve. For Pilots, there is a ship's wheel superimposed over an anchor in the center of the badge. For Assistant Marine Engineers and Marine Engineers, a ship's propeller is similarly superimposed. These are the only non-officer positions in the Department whose breast badge and cap badge are not identical.

Badges with the title Assistant Marine Engineer were discontinued in 1997. Anyone in this position is now issued Marine Engineer breast and cap badges.

The current Wiper's badge (for both breast and cap) is a standard Firefighter's Maltese cross. However, the central disk has an applied letter "W" preceding the Wiper's assigned badge number.

Marine Stoker badges.
(1891-c.1950 and c. 1950 to 1963)

Marine Wiper, formerly Stoker. (1963-Present)

Marine officer embroidered cap device.

A complement of firefighters is assigned to each marine company. They possess standard badges just like their *landlubber* counterparts. The cap devices of marine company officers are silver and display an anchor in addition to the one or two trumpets of rank. As previously described, company numbers have also been applied to officers' cap badges. Prior to May 1959, when fireboats were called "floating engine companies," the number on a Marine company officer's badge was that of the engine company number assignment. After May 1959, fireboats were each assigned a distinctive marine company number, which was the number appearing on the officer's cap badge.

Captain Leonard Gephard, circa 1891.

Marine officers' metal cap devices.
(1891-Present)

CHAPTER 8

Appointed Positions

The FDNY has varied personnel including uniformed members, "civilian" civil service staff, and appointed staff. The last category is led by and primarily reports to the Fire Commissioner. This includes the staff of Deputies and Assistants, as well as the Department's Counsel, Inspector General, and Secretary.

From the time of the Department's organization in 1737 until 1854, governance was by the Common Council of the City of New York. However, by 1854 the Council found that this task took up too much of its time, detracting from other City business. As a result, an ordinance was passed in 1854 for creating a "Board of Fire Department Commissioners." The Board was to consist of three representatives from each of the eight fire districts of the City plus three from the Twelfth Ward. The Twelfth Ward comprised everything north of 68th Street, which included the area above 86th Street where "suburban" volunteer companies continued to serve after the 1865 formation of the paid Department. The Chief Engineer was an ex-officio member of the Board.

FRANK LESLIE'S ILLUSTRATED NEWSPAPER

No. 517—Vol. XX.] NEW YORK, AUGUST 26, 1865. [Price 10 Cents.

The Atlantic Cable.

The unpleasant intelligence of the probable failure of the Atlantic cable has reached us, and again the world is disappointed; for all men, of every nation, were either directly or indirectly interested in the success of the enterprise. The last news from the steamship Great Eastern is received through the English papers of the 4th August. They report that on the 2d, 1,200 miles of cable had been payed out, when the signals became unintelligible. How this news was received through the line (and it could apparently have been received in no other way) is not very clear. It is probably only a conjecture made by the electricians at Toilhommerum cliff, and so telegraphed to the English journals as an explanation of the inevitable silence.

We learn from Valentia Bay, under date August 4, that no communication has been had with the ship since, and no information received. The cause was unknown, and at the closing of this summary nothing additional had transpired. The land lines between Valentia had been out of order.

The London *Times* infers that the disaster occurred after the cable left the vessel, from the fact that it was unknown to those on board, who were signalling their messages as usual, when their signals became unintelligible at Valentia, and at last ceased entirely. As to the

C. C. Pinckney. James W. Booth. — Chief-Engineer Kingsland. — P. W. Engs. Martin B Brown.

THE NEWLY-APPOINTED COMMISSIONERS OF THE PAID FIRE DEPARTMENT OF NEW YORK CITY.

The newly appointed Commissioners of the paid Fire Department of the City of New York as they appeared on the front page of Frank Leslie's Illustrated Newspaper, *August 26, 1865.*

The Commissioners underwent another change on May 29, 1855 when State legislation ordered that the Board consist of five members, all of whom would be elected by the members of the Fire Department. Additionally, qualifications for election were required. Candidates had to be Exempt Firemen who had not been active for a period of at least three years prior to their time of election. Initially, they were to be elected in staggered terms of one to five years. The intent was to replace one Commissioner in each annual election and all would eventually hold five-year terms. With this change, the Board of Fire Department Commissioners became the "Commissioners of the New York Fire Department." It was also in 1855 that badges were introduced for the Fire Department. There is no record as to whether a badge of any type was issued to Commissioners to show their status or to grant them access to the newly created "fire lines." It would seem necessary for them to have such access. But, no such badges were uncovered.

In 1865, after years of debate and criticism, the State passed legislation to create the Metropolitan Fire District and placed control of the Department into the hands of the State of New York. The legislation also specified that the Governor would nominate, and the State Senate would confirm, four persons to act as Commissioners. Terms were staggered initially but would eventually be for eight years. At this juncture there is evidence that badges were issued to the Fire Commissioners. The earliest Commissioner's badge still known to exist belonged to Joshua Grovesnor Abbe who served on the first Board of Commissioners of the Metropolitan Fire Department from it inception in 1865 until his death in September 1868. The badge is in the shape of a pinched shield and is made of fine silver. It is entirely hand-engraved, with an image of a steam fire engine. At the top is the name, "Metropolitan Fire Department Commissioner" and at the bottom is the name "J. G. Abbe."

Commissioner Joshua Grovesnor Abbe (1865-1868) and his custom, hand-engraved badge.

The badge of Commissioner Alexander Shaler, shown here, is reproduced from the landmark history book, *Our Firemen,* published in 1887. The Commissioner was referred to as General Shaler, not only for the rank he held in the Union Army during the Civil War, but also because he adopted military ranks for the Department in 1868. He served as Commissioner from 1867 through 1873. In 1874, General Shaler was retained as "Consulting Engineer" to the Chicago Fire Department to update and re-organize that department after its second devastating fire. General Shaler's badge is unique and unquestionably custom or jeweler-made.

Commissioner Alexander Shaler (1867-1873) and his custom-made, very unique badge.

The 1868 Annual Report of the Department describes a badge worn by Department "officials." It is not clear who these individuals were, but given that badges for specific commissioners have been documented, it may be that badges were also issued to other appointees. The badge for officials is described as being shield-shaped bearing the inscription, "Metropolitan Fire Department" with the monogrammed initials "M.F.D."

In 1870, the Commissioners experienced another change in status when control of the Department reverted back to the City and its name was changed to the Fire Department of the City of New York. Nomination of Commissioners became the purview of the Mayor. Around this time, another badge style was introduced into the Department for officers and appointed officials. Although badges with officers' ranks on them have been located, and are shown in Chapter 7, none with the title of Commissioner were located. However, one such badge issued to Joseph Perley is illustrated in Costello's, *Our Firemen.* Joseph Perley was appointed to the Board of Commissioners in May 1873 by Mayor Havemeyer for a six-year term, serving as President for four of those years.

Commissioner Joseph Perley (1873-1881) and his badge which probably represents a standardized badge for Commissioners and Officers.

The notion of a Board of Commissioners was abolished in 1898 when John Scanlon was appointed Fire Commissioner for the newly consolidated Greater City of New York.

Although they were never part of the uniformed force, Commissioners did, in fact, wear uniforms for a short period in the Department's history. When the Metropolitan Fire Department was formed, uniforms were first introduced for firemen and officers. Many reacted adversely to wearing them. So, to lead the way, the Commissioners donned uniforms. When this practice ceased is not known. But, it was likely to have been gradual and relatively unnoticed as firemen became more accustomed to wearing uniforms.

The only Commissioner's badge from the period 1884 to 1937 that has been located belonged to Deputy Commissioner P. A. Whitney. It is a pentagon-shaped badge with a banner added to the top in which the name is inscribed. It has been suggested that some Commissioners' pentagon badges may have had an eagle on top, but this has not been confirmed. The Commissioner is an appointed, non-uniformed "civilian" position. When the sun-ray Maltese cross was introduced in 1937 for uniformed officers, it may or may not have been issued to the Commissioner. A photograph of Commissioner Frank J. Quayle receiving his badge of office in 1941 clearly shows it to be a sun-ray Maltese cross.

This rare pentagon Deputy Commissioner's badge was used in the early 1900's but its provenance is uncertain.

Commissioner Quayle receiving his badge of office from Mayor O'Dwyer as Commissioner Murtagh and Chief Frank Murphy look on, January 1, 1946.

Three styles of Commissioner badges have been designed and manufactured since the inception of the sun-ray Maltese cross design. The standard sun-ray Maltese cross with eagle was used for the longest period. In 1971, stars were added to all Commissioner, Assistant Commissioner and Deputy Commissioner badges, just as they were for high uniformed ranks. The most recent badge style to be introduced was the police-style sunburst badge. This style badge was simultaneously used by the Deputy and Assistant Commissioners as well as for the Chief of Department and for staff Chiefs. As noted in Chapter 2, the representation of the Department's name on the police-style badges as, "New York City Fire Department" was another significant change that accompanied their issuance.

The police-style badge was instituted during the tenure of Commissioner Howard Safir who was later transferred by Mayor Rudolph Guiliani from Fire Commissioner to Police Commissioner. He is the third person in the history of the two departments to hold both positions. (Rhinelander Waldo, FC 1910-1911, PC 1911-1913; George P. Monaghan, FC 1951, PC 1951-1953; Howard Safir, FC 1994-1996, PC 1996-2000).

Variations of Commissioner's badge (1937-Present.)

Variations of First Deputy Commissioner's badge (1937-Present.)

Variations of Deputy Commissioner's badge (1937-Present.)

Custom First Assistant Commissioner's badge.

Miniature badge of Commissioner Patrick Walsh (1941-1945.)

This badge was ordered by Commissioner Thomas Von Essen in the style of the first standardized Commissioner's badge.

Badges in the Commissioner's Officer

Commissioner
1st Deputy Fire Commissioner
2nd Deputy Fire Commissioner*
Deputy Fire Commissioner
1st Assistant Fire Commissioner*
Assistant Commissioner
Secretary to Department
Department Counsel
Executive Assistant to Commissioner
Secretary to the Fire Commissioner**
Inspector General*
Confidential Secretary**
Special Investigator**
Special Assistant Fire Commissioner - Press**

* - Not currently used   ** - Not routinely issued

In 2001, Commissioner Thomas Von Essen proposed a new tradition in the FDNY. He ordered a badge to be made in the style believed to be like the first standardized Commissioner's badge in 1884. It will be a pentagon with eagle on top and made of solid gold. This will become the ceremonial "badge of office" to be handed from each Commissioner to his successor. Each Commissioner will continue to be issued a badge for identification in whatever the current standard design is at the time.

The Inspector General is the Department's internal affairs investigator. For many years, the person holding this position received a badge with an engraved title. Currently, an Assistant Commissioner badge is issued and the official title is Assistant Commissioner for Investigations.

There are many titles unique to the Commissioner's office which may come or go with the change of an administration. If badges are issued to these individuals, as they often are, they are typically custom-made and may vary from the standard two-piece, reverse-enamel ring design of the standard FDNY sun-ray Maltese cross badges. Although there are no written guidelines for such badges, several have been found and are shown here with direct-engraved lettering in red enamel.

Assorted positions appointed by the Commissioner.

Executive Assistant to Commissioner badge of Lt. C. Sansevero, assistant to Commissioner O'Hagan (1973-1978.)

There are two other categories of appointees that are very different from the ones described, namely Medical Officers and Chaplains. Unlike other appointed positions, which work mostly in the Department's executive administrative structure, Medical Officers and Chaplains fulfill services at the field level. For this reason, the information on their badges is included in Chapter 10 - Support Positions.

CHAPTER 9

Special Details

Five badges in the history of the Department have been issued for special details. Specifically, these were for Housewatch, Patrol, Theater Detail, Inspector of Licensed Places of Public Assembly and Photographer. Each badge was (or is) a temporary substitute for the uniformed member's standard badge to be worn only when performing the assigned duties. Of these, only one, Department Photographer, remains in use.

Illustration of a NYC Fireman on the newly instituted Patrol as appearing in Frank Leslie's Illustrated Newspaper *in January 1865.*

Patrol

The practice of sending a patrol out onto the streets of New York City every night to watch for, and respond quickly to fires, was initiated soon after the settlement of the colony in 1648. As the City grew, these patrols were eventually replaced by watchers in bell towers who sounded the alarm using a numbering system of gongs to alert firefighters and citizens to the location of the fire. When the paid Department was organized in 1865, the first set of Regulations were developed in September of that year. In them, the practice of a patrol or fire watch throughout the City was continued. The foreman of every company was required to maintain, *"a proper and efficient watch day and night, so that at all times two men shall be on patrol in the neighborhood of the engine or hook and ladder house, and one on watch in the house."*

In March 1869, changes were made to the patrol function. It was reorganized into three brigades and eight battalions. Strict rules for the patrol, as well as for the keeping of company journals, were also imposed at this time.

A badge was designed for use during patrol detail. It was large, round and slightly domed, bordered by a length of hose with a hydrant at the top. The badge was relatively large, measuring three inches high and two and one-half inches wide. This particular style Patrol badge was in use only during the MFD period from 1865 to 1870.

Patrol badge of the Metropolitan Fire Department (1865-1870.)

Two size variants of NYFD Patrol badge (1870 - ?)

The MFD Patrol badge was replaced, presumably in 1870, with a badge in the form of a cross though an equilateral cross, bearing the City seal in the center and the word, "Fire Department Patrol" in the concentric ring. Two versions of this badge exist; the principal difference is their size. The larger of the two measures three inches high, the same size as the MFD Patrol badge. This suggests that the larger version was the first of the two introduced, because it is similar in dimension to its immediate predecessor. The smaller version may have been designed in keeping with the size of the newly adopted Maltese cross or it may have been altered in January 1871 when specific "patrol beats" were designated throughout the City. In 1879, General Order No. 1 specified that the Patrol badge would be the current firefighter's Maltese cross, thereby ending the use of the distinctive patrol badge.

Very few patrol badges survive today. However, both distinct styles of patrol badges are shown above.

Engraving of a typical 19th century New York Fireman at Housewatch.

Housewatch

The 1865 Rules and Regulations also established the "Housewatch" as we continue to know it today. Some of the duties of the Housewatchman, as specified in the Regulations for the Uniformed Force in 1954, were as follows:

> Keep the fire zone in front of the quarters free from vehicles and obstructions.
>
> Promptly and neatly record the receipt of all alarm signals on the blackboard at the Housewatch desk.
>
> Be responsible for the condition of all apparatus in quarters and the sidewalk in front thereof.
>
> Protect the company journal and see that it is not defaced or tampered with.
>
> Record in the company journal all matters required by the Regulations, and be responsible for the accuracy of such entries.
>
> During night tours, promptly notify the officer of snowfalls which may necessitate the use of skid chains.
>
> Promptly sound the verbal alarm bell and report to the officer any visible fire, indication of fire, or receipt of verbal or telephone alarm.
>
> Attend to all heating apparatus, except when these duties are otherwise provided for.

The Housewatchman was required to wear his uniform cap with badge and, for some period of time, his jacket with breast badge as well. When an alarm came in for the company, the Housewatchman would remove his cap and make for the apparatus to don his firefighting gear. Unfortunately, many badges were stolen while the company was out of quarters. In an attempt to curtail the illegal possession of an FDNY badge, Commissioner Edward Cavanaugh decided to have a special "Housewatchman" badge designed and issued some time during the late 1950's.

Housewatch badge designed in 1950's but never issued.

The Housewatch badge is a simple oval with a high luster finish, bearing the inscription "Housewatchman" and the company number. However, after the badges were struck and mass produced, the idea was tabled. Around the same time, to avoid the penalty of losing their badge, many firemen took it upon themselves to order backup or "saver" badges (see Chapter 18). These saver badges were similar to the standard Maltese cross yet significantly different so they could be recognized as an unauthorized copy. Also, a simple wooden sign (about three inches x ten inches) with the fireman's name and badge number was used at the Housewatch desk for purposes of identification. When not on Housewatch duty, the sign would hang under the member's helmet and gear on the racks along the sides of the apparatus floor.

Housewatch badges remained in storage at the old Department Quartermaster at Pier A for many years. Around 1994, these badges were either discarded or appropriated out of the possession of the Department and into the hands of collectors.

Theater Detail

Broadway! New York is the very heart of live entertainment in the U.S., if not the world. The earliest accounts of theater performances in the City date back to about 1699. But, disasters were also a part of its flamboyant theater history. Theater fires were frequent and unfortunate occurrences, often causing catastrophic damage and loss of life. The most fatal New York fire occurred in 1876 at the Brooklyn Theater during a performance of *The Two Orphans*. Approximately 295 people died in the blaze. Many of the victims were unidentifiable and were laid to rest in a common grave in Greenwood Cemetery, where it can be seen today.

FRANK LESLIE'S ILLUSTRATED NEWSPAPER

No. 1,108—Vol. XLIII.] NEW YORK, DECEMBER 23, 1876. [Price, 10 Cents.

Holiday Number of Frank Leslie's POPULAR MONTHLY now ready. Price only 20 cents. An elegant present. 128 pp.—Over 100 Illustrations.

NEW YORK.—THE BURNING OF THE BROOKLYN THEATRE, DECEMBER 5th—SCENE FROM THE STAGE AT THE FIRST ALARM OF FIRE—THE ACTORS ENDEAVORING TO PREVENT A PANIC.—See Page 264.

Dramatic newspaper rendition of the terror in the Brooklyn Theater during the fire in 1876.

Chapter 742 of the laws of 1871 required that one or more firemen were to be present at all performances in licensed theaters in New York. It was the fireman's duty to monitor the theater for any activity that could cause a fire, to be alert to the occurrence of fire, to report it as quickly as possible, and to initiate appropriate firefighting or lifesaving. Chapter 742 of the laws of 1871 also made it illegal to falsely shout *"FIRE!!"* in a crowded theater.

In 1882, Commissioner John Gorman sent a report to the Board recommending enactment of strict fire codes for theaters. These codes were revolutionary at the time and were seen as a major step forward in preventing the loss of life that typically accompanied theater fires.

General Order No. 4, issued May 20, 1884 described and authorized the use of a badge to be worn during assignment to theater detail. (A copy of the full text of the Order is contained in the Appendix.) The old English spelling, "theatre" appears on the badge. This badge was in continual use until 1960.

Standard Theater Detail badge (1884-1960)

Specifications for the Theater Detail badge in the 1926 Rules & Regulations of the Department indicate that the enamel of the lettering on the badge is to be blue. However, one badge was found with red enamel lettering. This unusual variation cannot be explained. The badge appears to be genuine in every other respect. A customized theater detail badge was found with a fireman's name engraved in place of a number. Reportedly, this fireman was a frequent volunteer for theater detail.

Unusual variations of the Theater Detail badge.

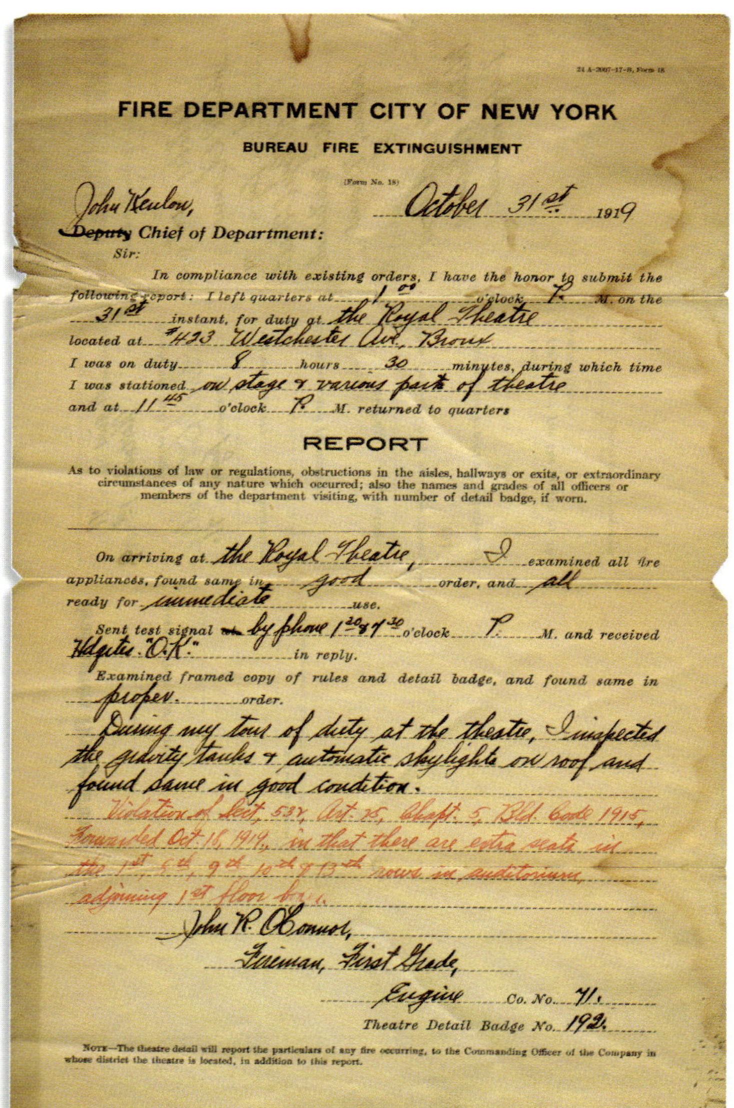

24 A-2007-17-B, Form 18

FIRE DEPARTMENT CITY OF NEW YORK

BUREAU FIRE EXTINGUISHMENT

(Form No. 18)

John Kenlon, October 31st 1919

~~Deputy~~ Chief of Department:

Sir:

In compliance with existing orders, I have the honor to submit the following report: I left quarters at 1^{00} o'clock P. M. on the 31st instant, for duty at the Royal Theatre located at #423 Westchester Ave. Bronx

I was on duty 8 hours 30 minutes, during which time I was stationed on stage & various parts of theatre and at 11^{45} o'clock P. M. returned to quarters

REPORT

As to violations of law or regulations, obstructions in the aisles, hallways or exits, or extraordinary circumstances of any nature which occurred; also the names and grades of all officers or members of the department visiting, with number of detail badge, if worn.

On arriving at the Royal Theatre, I examined all fire appliances, found same in good order, and all ready for immediate use.

Sent test signal by phone 1^{20} & 1^{30} o'clock P. M. and received Hdqtrs. "O.K." in reply.

Examined framed copy of rules and detail badge, and found same in proper order.

During my tour of duty at the theatre, I inspected the gravity tanks & automatic skylights on roof and found same in good condition.

Violation of Sect. 537, Art. 25, Chapt. 5, Bld. Code 1915, Forwarded Oct. 18, 1919, in that there are extra seats in the 1st, 5th, 9th, 10th & 13th rows in auditorium adjoining 1st floor box.

John R. O'Connor,
Fireman, First Grade,
Engine Co. No. 71.
Theatre Detail Badge No. 192.

NOTE—The theatre detail will report the particulars of any fire occurring, to the Commanding Officer of the Company in whose district the theatre is located, in addition to this report.

Copy of a Theater Detail Report dated October 31,1919 for the Royal Theater in the Bronx.

A very nice fictional story about a fireman on theater detail falling in love with the star of the play, written by Dr. Robert Holzman, (then a member of the NYFD Auxiliary Corps and later author of the historic book, *The Romance of Firefighting*) can be found in the April 1942 issue of *WNYF*.

Inspector of Licensed Places of Public Assembly

The Division of Licensed Places of Public Assembly was responsible for, *"enforcing all laws and Rules of the Board of Standards and Appeals in relation to the protection against fire and panic, obstruction of aisles, passageways and means of egress, standees, fire prevention and fire extinguishing appliances, and fire prevention in all licensed places of public assembly."*

Modified Theater Detail badge for Inspector of Licensed Places of Public Assembly (1920-1960.)

The title of "Inspector of Licensed Places of Public Assembly," is rather long and difficult to fit onto any badge. However, such a badge was developed and issued from 1920 through 1960. The badge was a theater detail badge that was modified by soldering on an engraved panel with the title on it. It was done to make use of surplus theatre detail badges that had been struck. These badges also provided some uniformity since uniformed officers who were Inspectors of Licensed Places of Public Assembly were responsible for supervising firemen on theater detail duty.

Although this detail no longer exists, certain places of public assembly within New York City pose serious threats to life and property. Such places as "social clubs" have become targets for enforcement of fire codes particularly after the Happy Land Social Club fire in 1990 which claimed 87 lives. Two current positions related to this activity, "Enforcement Agent" and "Process Server" are civilian support positions included in Chapter 10.

Member of the FDNY Photo Unit at the New York Telephone Company switching center fire in 1975.

Photographer

The only assignment for which an official badge is currently issued is Department Photographer. When assigned, a sun-ray style Maltese cross with the title "Photographer" in the ring is substituted for the standard Department badge with number. The Photographer badge is issued regardless of rank. Firemen assigned to photographer duty return their original badge. However, when the breast and cap badges are turned in to the Badge Desk, they are filed in the envelope of the Photographer badge that is assigned to them.

The Department Photographers play an important role in fire investigation, working hand in hand with Fire Marshals. The badge establishes the photographer's identity and role at the scene.

FDNY Photographer working with fire marshals to document investigation of a suspicious fire.

Department Photographer (1937-Present.)

CHAPTER 10

Non-firefighting Support Positions

A standard badge for non-firefighting support personnel has been issued since 1884. Over the years, there have also been numerous badges issued to "civilian" employees. In 1996, Deputy Commissioner Feehan issued an order to eliminate the use of badges by anyone but uniformed members of the Department. So, this tradition is coming to an end.

The badges in this Chapter have been divided into two categories, namely civilian and uniformed. These two categories can be distinguished by their badge color. Civilian badges are silver. Badges for uniformed personnel are gold. The one exception to this is that high-level appointed civilian support positions in the Commissioner's office and supervisory civilian positions, are issued gold-color badges.

The term, "uniformed non-firefighting support positions" refers to positions that are either promotions from the ranks of the uniformed force or for which a rank in the uniformed force is granted. For example, all Fire Marshals are appointed from the uniformed force. They are not required to wear a uniform and, in fact, no cap device is issued to them. Medical Officers and Chaplains are appointed from members of the professional community and, upon appointment, they are given a uniformed rank in the Department. Historically, Medical Officers and Chaplains have worn uniforms. Cap devices are issued to them for this purpose. Chaplains continue this tradition while, in recent years, Medical Officers have moved away from it.

Civilian Positions

Through the years there has been, and continues to be, a long list of civilian positions within the Department, many of which are issued a badge. The first record of badges being issued to civilians was in 1884. General Order No. 4, dated May 20, 1884 gave specifications for the pentagon-shaped badge. This badge shape continued to be used for some civilian employees into the 1950's until the Maltese cross with eagle was issued.

Over the years, civilian positions and their titles have come and gone. There has also been some consolidation of badges within groups, such as "Communications," "Accounts and Procurement," and "Transportation."

In March 1990, the Happy Land Social Club fire claimed the lives of eighty-seven people. The Club was found to have numerous fire code violations that impeded escape from the building and increased fatalities. To crack down on such illegal clubs, the Department instituted the position of "Process Server" to issue subpoenas and other legal documents to the owners of such establishments. A new style badge, different in appearance from other FDNY badges, was designed for this position. Similar to NYPD badges, the Process Server badge is a sunburst style with a ring bearing the words, "Fire Department City of New York"

Unusual badge design introduced for Process Server in 1991.

Pentagon badges for civilian positions, first issued in 1884 and continuing at least into the late 1940's.

Defective badge, improperly trimmed, held by the Department and never issued.

Unusual, but verified, Department issued Storekeeper badge.

and a panel at the bottom with the title "Process Server." Currently, this is the only support position badge in a sunburst shape. The sunburst, however, was also introduced for the Department's top uniformed and appointed positions (see Chapters 7 and 8).

The issuance of badges to most civilian positions is primarily for status. With few exceptions like Inspectors or Process Servers, it is unlikely that civilian employees of the Department would ever find themselves in a practical situation where they would need to display this symbol of authority. Rarely is the badge ever displayed on the clothing. Civilian support positions have been issued a breast badge only with no cap device.

Civilian badges from the horse-era (1884-1922.)

During the horse drawn era of the FDNY, numerous positions were required for the care of these venerable equine members of the Department. Civilian badges issued during this period included such positions as Superintendent of Horses, Blacksmith, Hostler, and Harness-maker. These positions fell within the Division of Horses.

While the function of some of these positions is obvious, many readers might not know that a "Hostler" was the person who ensured the proper care of the horses throughout the Department. In 1906, the first horse ambulance was placed in service. In 1910, ten horse-shoeing trucks were established. The last horses used to power fire-fighting apparatus made their final run in 1922. As horses were phased out of service over the next three years, so were their support positions and badges. The horse drawn era, to say the least, was a very colorful period in the Department's history.

Steam fire engines used between 1858 and 1933 required that the boiler be maintained at a temperature and pressure level sufficient for pumping at the scene of the fire. Boiler maintenance was assigned to a civilian Stoker. Typically, the boiler for heating the firehouse was connected to the steamer to keep it at a standby temperature and pressure. Quick-disconnect valves were installed so that the engine could be quickly and easily underway. In some departments, this process was reversed and the steamer provided heat for the firehouse. Once at a fire, it was the Stoker's job to keep the boiler fire hot. He worked hand-in-hand with the Engineer. In the 1865-1866 Annual Report of the Metropolitan Fire Department, the Stoker's job was described as follows:

> *"The stoker shall, under the direction of the engineer of steamer, assist in the care and cleanliness of the engine; they shall see that the engine is properly supplied with fuel, and perform such other duties as the foremen prescribe."*

FDNY Blacksmith wagon circa 1910. (Photo courtesy of John A. Calderone.)

Stoker's badge from the steamer-era (1884-1933.)

Stoker manning the steamer of Engine Co. 5.

The Stoker clearly had an important role both in the firehouse and at the scene. It is also interesting to note that in other "industries," the title of the person who fulfilled this role was often called a "Fireman." Clearly, the Department had to come up with the alternative title of Stoker.

As described in Chapter 7, the civilian support position entitled Marine Stoker became the uniformed position Marine Wiper in 1963. Initially, it was the job of the Marine Stoker, like that of steamer Stoker, to maintain fire in the boiler. Now the Marine Wiper has responsibility for all below-deck operations aboard the fleet of fireboats operated by the Department.

A forgotten, but important support position was that of Matron. Matrons were traditionally the widows of firemen who died outside the line of duty, and therefore did not receive death benefits. The women served as "House Mothers" of fire companies. The first Matron, then called a *janitrix*, was Mary Sanders, hired in 1873 to clean Headquarters at a salary of $300 per year. In 1944 there were 550 Matrons on the job. In 1954, the monthly salary for Matrons was $75. Their tasks included the typical housekeeping duties of the day including laundry, cleaning, cooking, etc. The task of scrubbing floors was removed from the Matrons by Chief Edward Croker who reassigned this to the firemen, believing it was too strenuous for the women. Matrons were paid by the men in the company not by the Department. The Department's role was merely one of tacit approval and oversight. Pentagon badges were typically issued to Matrons. The last Matron retired in 1993. She was Grace Faessler, the widow of Lt. Jacob Faessler.

Miscellaneous civilian positions (post 1950.)

The Central Fire Alarm Office, established in 1853, is known today as the Communications Division. It is shown here in a photograph circa 1910. (Photo courtesy of Steven Scher.)

Early civilian positions of the Communications Division.

Since the earliest days of the Department, dedicated staff transmitted alarms. The earliest organized approach to this important function began with the "rattle watch" in 1658. Since then there have been significant milestones in the history of fire communications in New York City, chronicled as follows: construction of fire bell towers in 1845, assignment of paid Bellringers and Assistant Bell ringers in 1853, the introduction of the telegraph in 1851, the establishment of a central fire alarm office in 1853, installation of street fire alarm boxes in 1873, establishment of the Bureau of Fire Alarm Telegraph in 1911, installation of two-way radio telephones in the first firefighting apparatus (fireboats) in 1937 and all other fire apparatus in 1945, notification of alarms by voice instead of bells in 1969, connection to the 911 telephone emergency number in 1972, and the computer-aided "StarFire" dispatch installed in 1980.

Communication Division positions circa 1950 to present.

One-of-a-kind, Department-issued, Communication Division badges.

Throughout this history there have been numerous civilian and uniformed support positions for fire alarms and communications. Most of these were given individual badges until the late 1970's when all were replaced with one badge bearing the unit name, "Communications."

Since the appointment of Fire Wardens in 1648, the Fire Department has played a role in inspecting various potential fire hazards within the City, enforcing fire codes, and imposing citations. Inspectors were typically assigned a specialty such as combustibles, blasting, electrical, etc. Like so many other support staff, many of these titles and their badges have come and gone. At the present time, badges related to these activities are issued under two titles namely, Inspector and Enforcement Agent.

Inspection-related pentagon badges (1884 to around 1950.)

Inspection-related positions after 1950.

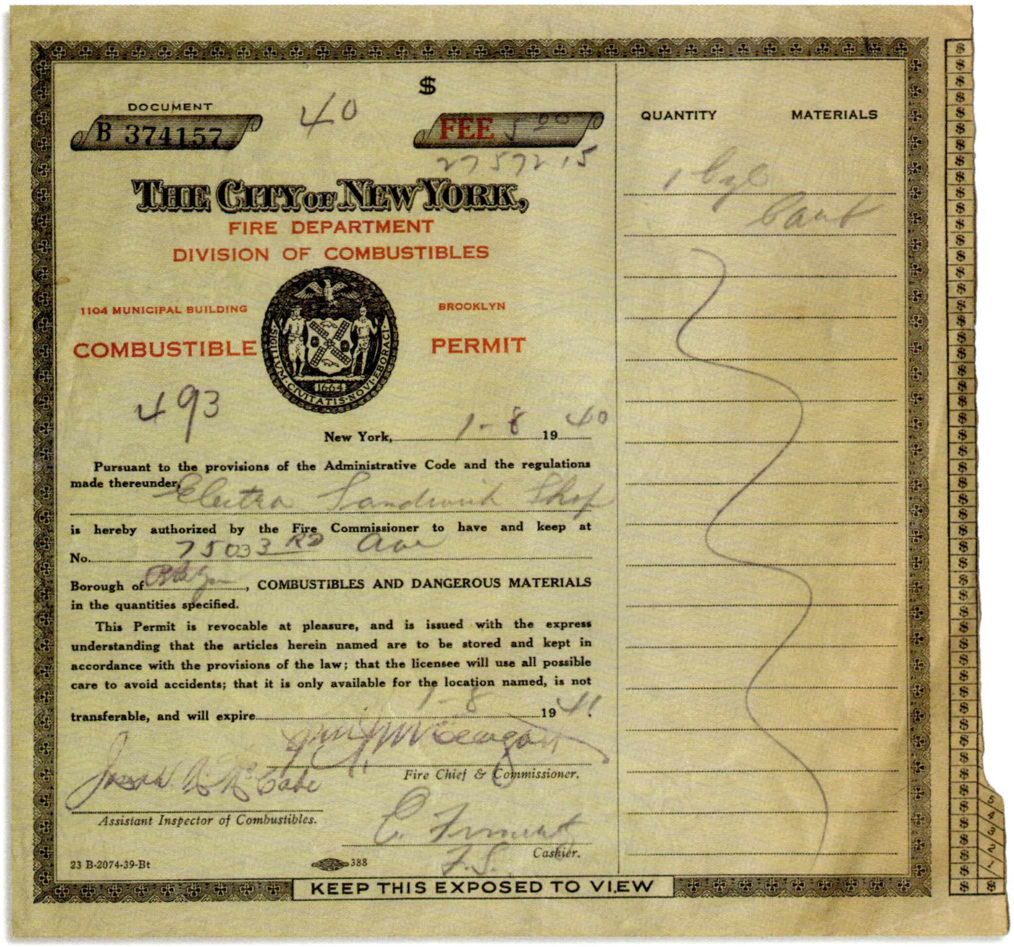

DOCUMENT B 374157 40 $ FEE

THE CITY OF NEW YORK,
FIRE DEPARTMENT
DIVISION OF COMBUSTIBLES

1104 MUNICIPAL BUILDING BROOKLYN

COMBUSTIBLE PERMIT

493

New York, 1-8 1940

Pursuant to the provisions of the Administrative Code and the regulations made thereunder Electra Sandwich Shop is hereby authorized by the Fire Commissioner to have and keep at No. 7503 3rd Ave Borough of Bklyn, COMBUSTIBLES AND DANGEROUS MATERIALS in the quantities specified.

This Permit is revocable at pleasure, and is issued with the express understanding that the articles herein named are to be stored and kept in accordance with the provisions of the law; that the licensee will use all possible care to avoid accidents; that it is only available for the location named, is not transferable, and will expire 1-8 1941

Fire Chief & Commissioner.

Assistant Inspector of Combustibles.

Cashier.

23 B-2074-39-Bt 388

KEEP THIS EXPOSED TO VIEW

QUANTITY	MATERIALS
1 Gal	Paint

Certificate issued by an Assistant Inspector of Combustibles for storage of combustible materials in 1940.

Current Division of Repairs and Transportation badges.

A view of the Department Shops in Long Island City in 1948.

The Department has been mobile since delivery of the first wheeled apparatus to the FDNY in 1731. Over time, this mobility has been achieved by manpower, horsepower, and steam, gas or diesel engine. The Department established its Repair Yard at Elizabeth Street in Manhattan upon formation of the Metropolitan Fire Department. Currently, the Division of Repairs and Transportation is housed in the approximately 112,000 square foot "Shops" in Long Island City. The Division has had an ever-changing staff to provide the essential service of maintenance. Many interesting titles are found on the badges of the civilian staff of the Division, reflecting the Department's history. They have included such civilian support positions as Auto Mechanic, Rubber Tire Repairer (rubber tires were introduced on Department apparatus in 1899), and Wheelwright. In 1947, all civilian badges within the Bureau were replaced with the Maltese cross with eagle badge bearing the unit title, "Transportation." Badges ceased to be issued to members of this Division in the early 1980's.

Accounts & Procurement badge given to headquarters support staff.

Civilian-employee badge issued to Public Information Officers.

Since the earliest days of the Department, many administrative functions have been performed by civilian staff. In fact, many historians have overlooked the fact that paid staff within the Department were present long before firefighters were first paid in 1865. Paid staff included Bellringers, Messengers, and Commissioners. Badges displaying job titles have been issued to administrative and clerical positions. Most badges with job titles have now been consolidated under the broad badge title, "Accounts and Procurement."

Although civilians, several positions are appointed and generally work within the Commissioner's office. The badges for these positions are included in Chapter 8. All badges in this category are gold in color. On occasion, at the request of the Commissioner, individual badges have been made for members of the Commissioner's staff. Since such badges may be issued only once and since there are no existing strike plates for such titles, these badges may appear significantly different than the standard FDNY Maltese cross with eagle badge. Often they have the title and Department name engraved and enameled directly or may have them in reverse enamel style. In either case, there is no applied ring. Some collectors presume that these badges without an applied ring are fake. However, this may not be the case.

Uniformed Positions

In the spring of 1854, a number of suspicious fires caught the attention of Alfred E. Baker who brought them to the attention of the police. In response, the police justices appointed him to be their clerk and directed him to investigate the fires. He was not paid by the police department for this service, so the insurance companies raised funds to pay him. They also bestowed upon him the title of Fire Marshal. He was very effective in his duties and a year later the insurance companies increased their financial support and appointed an assistant for him. The Board of Police Commissioners approved and presented Baker with a badge, in the form of a police sergeant's shield, with the title "Fire Marshal, New York" engraved upon it. NYFD Chief Engineer Carson approved

Alfred E. Baker, first New York City Fire Marshal, with a facsimile of the badge issued to him by the New York Police Department in 1855.

Alfred Baker's request to wear fireman's gear including a red shirt, helmet and fire coat. However, he still reported to the Board of Police Commissioners. In 1868, the office of Metropolitan Fire Marshal was created by law, remaining under the jurisdiction of the police department until another statutory change in 1871 when the Metropolitan District was abolished. At this time, the name of the fire department was also changed and the title Fire Marshal was changed to City Fire Marshal. In 1873, the Bureau of the Fire Marshal was moved to the jurisdiction of the Fire Department and appointments to the position were made by the Fire Commissioners. The first Fire Marshal in the FDNY was George H. Sheldon.

The position of Fire Marshal was transferred to the NYFD in 1873. George H. Sheldon held the first appointment.

Bureau of Fire Investigation badges (1884-Present.)

Within the current Bureau of Fire Investigation there are several levels of Fire Marshal ranks with corresponding badges. These include Chief Fire Marshal, Deputy Chief Fire Marshal, Supervising Fire Marshal and Fire Marshal.

Other titles, such as Assistant Fire Marshal, were used during the civilian era and may be found on old pentagon badges. Fire Marshals were civilian appointees until the early 1960's. Currently, these appointments are only made

Unique, custom badge for "Counsel" Fire Marshal.

Contemporary Fire Marshals consulting with the Chief in charge at the scene of a suspicious fire. (Photo courtesy of FDNY/Fire Safety Education Fund)

Assistant Fire Marshal pentagon badge (1884-1937.)

from the ranks of the uniformed force. Bureau of Fire Investigation badges are all of the sun-ray Maltese cross with eagle design. In addition to being experienced firefighters and officers, Fire Marshals hold the distinction of being sworn police officers in the State of New York.

Physicians and dentists are appointed to the Department and are assigned to the Bureau of Health Services, previously known as the Medical Division. The first Department Surgeon, Dr. Charles B. McMillan, was appointed in November 1867. The "Medical Force" of the Department, established in February 1881, included one Medical Officer and two Vice Medical Officers. As of May 21, 1904, Medical Officers have been inducted into the uniformed force with a rank equivalent to their position. For example, a Medical Officer is equivalent to a Battalion Chief, while the Chief Medical Officer is equivalent to a Deputy Chief. Administratively, Medical Officers report to the Chief Medical Officer who in turn reports directly to the Commissioner. Medical Officers are on-call on a rotating schedule, for which they are assigned a marked car and driver.

Medical Officers do not commonly wear uniforms although some from the early days in the Department's history did wear them. Therefore, in addition to the standard breast or wallet badge, two distinct cap badges exist for the two Medical Officer positions, consisting of either two or three crossed trumpets with a red enamel equilateral cross superimposed.

The role of Chaplain is to give spiritual guidance and counseling as needed by members of the Department. This vital position has been a part of the Department since 1899. Rev. Father William St. Elmo Smith and Reverend James LeBaron Johnson were the first Department Chaplains. The March 28, 1899 *Transactions of the Department* state that they were given the rank of Chief of Battalion without compensation. Like Medical Officers, this rank continues to be conveyed upon Department Chaplains. The number of chaplains has varied over the Department's history. Typically about six chaplains represent various faiths or denominations.

Chaplains are entitled to and have traditionally worn a uniform. Cap devices displaying three crossed trumpets are designed for Chaplains. The trumpets are topped by either a crucifix or a Star-of-David, symbolizing either the Christian or Jewish faith of the Chaplain. There are no other sectarian cap devices.

Medical Officer and Chief Medical Officer breast badges (1937-Present.)

Custom Medical Officer badge of Dr. Lewin circa 1940.

Medical Officer cap devices.

Chaplain breast badge (1937-Present)

Chaplain cap devices.

Chaplain Fred P. Eckhardt at a working fire circa 1967.

Honorary positions exist for Medical Officers and Chaplains. Often these Honorary Medical Officers or Chaplains play an active role in the Department on a volunteer basis. In some cases, Honorary Officers have gone on to receive full appointment to the Department. Ironically, the Department's most well-known Honorary Medical Officer Dr. Harry M. Archer, never received an appointment as a Medical Officer.

Dr. Archer's love for the FDNY began in 1886 when, as a young medical student, he traveled to Europe with Commissioner Henry Purroy. He was not only a devoted Medical Officer from 1894 through 1940, he was an avid Buff, often sleeping in quarters. At the request of Mayor George McClellan, he traveled with members of the Department on it longest run to the Great Fire at Baltimore in 1904. Dr. Archer refused payment for his services to the Department. Therefore, he retained an Honorary title until appointed to the position of Second Deputy Fire Commissioner in 1940. Dr. Archer was appointed Honorary Battalion Chief in March 1907 and promoted to Honorary Deputy Chief in 1910. A fireboat was named after him in 1958, and in 1917 a medal was endowed in his name. Dr. Archer was awarded the Department's highest honor, the James Gordon Bennett medal, in 1924.

All non-firefighting positions on fireboat crews were initially civilian but have since become part of the uniformed force. These positions include Engineer, Assistant Engineer, Pilot, and Marine Stoker/Wiper. Although these are support positions, their current and historic badges are presented in Chapter 7.

Badge Titles For Civilian Support Positions

Blasting Inspector
Inspector of Blasting
Supervising Blasting Inspector
Inspector Fire Protection
Associate Fire Protection Inspector
Bureau of Fire Dept Buildings
Bureau of Buildings
Chief Inspector*
Deputy Chief Inspector*
Inspector*
Supervising Inspector*
Chief Inspector Electrical Appliances
Inspector of Buildings
Inspector of Electrical
Executive Inspector
Asst Inspector of Combustibles
Communications
Deputy Director Dispatch Operations
Supervising Fire Alarm Dispatcher
Fire Alarm Dispatcher
Radio Operator
Telephone Operator
Communications Electrician
Supervising Communication Electrician
Accounts & Procurement*
Public Information Officers
Engineer
Assistant Chemical Engineer
Assistant Civil Engineer
Assistant Mechanical Engineer
Community Relations*
Chief Instructor
Director Discipline
Electrician
Supervisor Electrical Installation
Supervising Electrician
Plumber
Supervising Plumber
Steamfitter
Supervisor of Mechanics
Machinist's Helper
Auto Mechanic
Rubber Tire Repairer
Transportation
Motor Vehicle Operator
Chauffeur
Clerk
Attendant
Foreman of Blacksmiths
Stoker
Messenger
Draughtsman
Telegraph
Wrecking Crew
Special Investigator
Assistant Attorney
Hostler
Foreman - Hostler
Asst. Engineer - Bureau of Fire Prevention
Telegraph Lineman
Superintendent of Horses
Watchman
Wheelwright
Process Server*
Matron
Enforcement Agent*
Chief Auditor
Safety & Health Coordinator
Industrial Hygienist
Storekeeper
Locksmith*

* - Badges in use in 1999

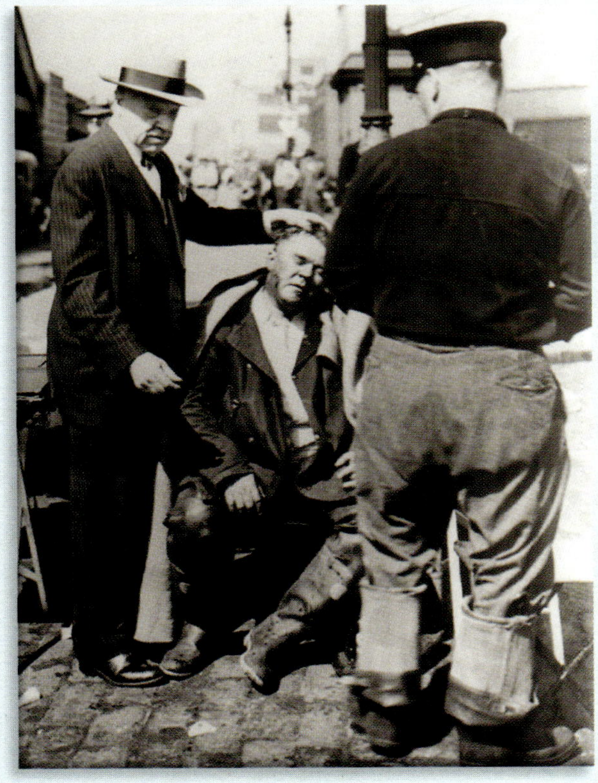

Honorary Medical Officer Dr. Harry M. Archer tending to injured firefighter.

Badge Titles For Uniformed Support Positions

Chief Fire Marshal
Deputy Chief Fire Marshal
Assistant Chief Fire Marshal
Supervising Fire Marshal
Fire Marshal
Assistant Fire Marshal*
Counsel - Fire Marshal
Chief Medical Officer
Deputy Chief Medical Officer
Medical Officer
Chaplain
Assistant Chaplin*

* - Obsolete

CHAPTER 11

Special Units

Early photograph of the FDNY band circa 1958.

In the course of the FDNY's history, numerous special units have been formed but only a few of them were issued special badges. Only one of the special units represented in this Chapter was comprised of uniformed firefighters, although all were officially part of the Department.

FDNY Band and Glee Club

The first Department Band was formed on August 10, 1900. It failed early on but was revived in 1913 and eventually had eighty-five members at its peak. The Band was made up of fireman who participated in this extracurricular activity due to their musical interest and abilities. The Band performed at many Department and City functions and marched in numerous parades along with other members of the Department. When performing, the members wore their regulation uniform modified by a blue cap with silver strap and a distinct badge. The badge was a regulation FDNY Maltese cross with a raised lyre, a harp-like instrument, in the center. The Band was dis "banded" on June 1, 1958.

Cap and badge of the Department Band (1913-1959.)

Badges from the music pouch of FDNY band members.

Several variations of the badge were found with "BAND" in letters affixed below the lower arm. This badge was not worn on the uniform. It was affixed to a pouch in which music was carried. The word, "BAND" was die-struck into the badge, or made of cut-out letters applied to the badge. Badges with cut-out letters were either nickel or gold-plated.

Badge from the Bugle & Drum Corps of the Department band. Date unknown.

Music pouch with badge.

One badge with a plate added to the bottom arm displaying the engraved letters, "B. & D. Corps" raises questions. No documents describing a bugle and drum corps (B. & D. Corps) in the Department could be found. Yet, this badge is in the possession of a descendent of an early band member. This leads to the assumption that the badge is authentic and that the Department band, like most others of the era, did have a bugle and drum corps.

The band was not the only musical unit of the Department. In early 1925, a Glee Club was formed as an activity of the Holy Name Society of Brooklyn and Queens, a Department line organization (see Chapter 15). However, the Glee Club soon grew into a Department-wide activity. No officers were permitted to be members of the group. Unfortunately, the Glee Club succumbed one year before the band in 1957.

The FDNY Glee Club rehearsing in 1954.

Badges of the Department Glee Club (1925-1957.)

Several badges, similar in style to the band badge, are believed to be associated with the Glee Club. One is a small Maltese cross with a lyre in the center on a field of red enamel. Like the band badges, it has a screw-post attachment suggesting that it was worn on a hat. Another is a smaller Maltese cross, only about one inch in diameter, with the lyre and red enamel background. This may have been a lapel badge.

Fire Auxiliary Corps

At three times during its history, the FDNY was supplemented by an auxiliary corps. In two instances, world wars were the impetus for the creation of the auxiliary. The third instance was prompted by the onset of the Cold War. Under such conditions, it was reasonable to expect that the ranks of the uniformed force could be depleted by the need of the armed forces for able-bodied men. Enemy attacks within the City requiring extraordinary firefighting resources beyond the abilities of the Department were also a possibility.

The first auxiliary was organized in 1918. Initially, the Department planned to hire temporary firemen to enhance any depletion of the uniformed force by the war. However, legal battles delayed the program. The concept of a "Volunteer Auxiliary Corps" was successfully introduced. Three thousand seven hundred men volunteered during the first six months of operation. Although not originally formed to perform firefighting duty, the volunteer corps was soon doing so. On November 5, 1918 the first full company staffed solely by auxiliary firemen was assigned its own apparatus and went into service.

NYFD Auxiliary Corps breast badges (1918-1919.)

The Auxiliary Corps had its own headquarters at 153 West 68th Street with this Auxiliary Corps company stationed there.

1918-1919 Auxiliary Corps Chief's cap with standard device.

A unique badge was designed for members of the Auxiliary. It consists of a keystone-shaped shield with rounded top. A section of hose serves as a border containing the raised letters "N.Y.F.D." and a fire helmet rests on top. The words "Volunteer Emergency Service" are present in raised letters across the central part of the shield and the member's number is applied toward the bottom. The badges were made of nickel. The basic badge was modified for Auxiliary officers by the addition of a tab attached to the base of the badge noting the officer's rank (e.g., Sergeant, Lieutenant, Captain, District Chief and Chief). Sergeants were elected in any company that had four or more auxiliarists. Lieutenants and Captains were selected along FDNY Battalion lines. There was one Chief per borough, selected and appointed by the Fire Commissioner. The Chiefs were all prominent businessmen and community leaders.

The *General Rules* established for the Auxiliary were brief in their reference to the badges, stating only the following:

> *"Section 32. No member shall dispose of such badge and loaning of same to any person shall be cause for dismissal.*
>
> *Should any badge be lost, the loss shall at once be reported to Sergeants and Commanding Officers of the Companies.*
>
> *When ceasing to become members of the Auxiliary Corps, badges together with all equipment that may be in possession of the member shall be returned to the Commanding Officer of the Company.*
>
> *When reporting in at fires members shall display badges and hold themselves in readiness to perform any duty that may be assigned to them."*

A uniform was authorized for members of the Auxiliary Corps consisting of a coat, khaki pants and a regulation FDNY hat with a red cord. A small Maltese cross was affixed to the hat that read "Auxiliary Corps" in the center with a pebbled background. No number or rank was affixed to the cap badge. Officers wore a standard rank-type cap device.

William J. Daly Auxiliary Corps Chief for the Bronx 1918-1919.

1918-1919 NYFD Auxiliary Corps cap and cap device.

The short-lived Auxiliary Corps was disbanded on May 1, 1919. At the conclusion of service, a "Badge of Honor" was bestowed on former members. It was a lapel pin in the shape of a Maltese cross with a blue enamel ring bearing the words, "Auxiliary Corps War Fireman." Displayed in the center is a very beautiful image of a fireman carrying a child down a ladder with a burning building in the background.

Badge of Honor" presented to discharged members of the WWI Auxiliary Corps.

The second time the Auxiliary was activated was in 1941 by Mayor LaGuardia, National Director of Civil Defense and avid FDNY Buff . This time, the fear of an attack on the continental United States was heightened by the devastating attack on Pearl Harbor and, in 1942, Mayor LaGuardia stated that New York City was a primary target of the Axis. During World War II, the role of the "Emergency Auxiliary Corps" (EAC) was to supplement the uniformed force called to duty, but to also prepare for fires that might result from attacks on the city. At its height, the EAC reached more than 55,000, nearly fifteen times the number during World War I and more than five times larger than the uniformed force at the time. During this period, the auxiliarists' role in firefighting was routine and the Corps was an integral component of fire protection in New York City. The Corps was disbanded in June 1945.

Due to the shortage of metals during World War II, metal badges were not developed or issued to the Auxiliary Corps. Rather, they were given wallet-size identification cards. Auxiliarists wore cloth armbands bearing the Civil Defense Fireman insignia. The insignia was also placed as a decal on the red "steel pot" helmets worn at fires.

Aside from supplementing the fire fighting force of the FDNY, a distinct auxiliary program was established for communications support. Several interesting Auxiliary Dispatcher badges have been found. The only official Auxiliary Dispatcher badge known to have been issued by the Department is not a badge by most standards. It is a celluloid pin-type badge. It is three inches round with the words "Auxiliary Dispatcher" in black on a field of yellow/orange. These badges were issued in the World War II Emergency Auxiliary Corps dispatcher program. Two unofficial Auxiliary Dispatcher badges are shown in Chapter 18.

Celluloid badge of Auxiliary Corps dispatcher; the only official Auxiliary Corps badge issued during the World War II period.

Fear of a nuclear attack on New York City was palpable during the Cold War. In 1950, an FDNY Auxiliary was mustered once again in response to a significant nationwide civil defense movement. The Corps was outfitted with 65 Ward LaFrance pumpers purchased with fed-

WWII Fire Auxiliary Corps practicing with a portable pumping apparatus in 1942.

Auxiliary dispatchers of the Brooklyn Emergency Operations Center in 1945.

eral civilian defense funds. Equipment provided to the Auxiliarists included standard Cairns New York helmets, painted yellow to distinguish them from the paid force. The frontpiece bore the number of the company to which the auxiliarist was assigned, with "AUX" displayed where the badge number is normally embossed. The badge designed for the Auxiliary consisted of a standard Maltese cross with an enameled Civil Defense logo in the center, the letters "Aux. FDNY" engraved in black enamel in a flat panel in the upper arm and the company assignment in a similar panel in the lower arm. The member's number was affixed to the rear of the badge, inscribed into a soldered-on disk.

Department-issued post-WWII Auxiliary Corps badge with number plate on back.

The Fire Auxiliary Corps is still in existence today. However, its ranks are significantly less than during the war periods. Badges and other supplies are not provided by the Department. Members are only issued standardized identification cards. If the threat of war once again poses a need, the Auxiliary could conceivably be re-activated to its historical status.

Non-department issue Auxiliary Corps Captain badge still in use.

Fire Salvage Corps

The Fire Salvage Corps was formed in 1972. It should not to be confused with the New York Fire Patrol, which has also been referred to and known as the Salvage Corps at various times in its history. The Fire Salvage Corps was one of four special projects established with funding from the U.S. Department of Housing and Urban Development's, "Model Cities" program. The functions of the Salvage Corps were similar to those of the Fire Patrol. However, their coverage was limited to the areas

Members of the FDNY Salvage Corps in 1972 distributing literature on Christmas tree fire safety.

of the city participating in the Model Cities program. This was initially Central Brooklyn and later included other areas of the City.

Membership in the Salvage Corps, a federal project to enhance life in urban areas, was open only to young males ages nineteen through twenty-seven who lived within the community covered by the program. In addition, medical and physical fitness examinations were required and anyone with a record of drug abuse, conviction of a felony, or dishonorable discharge from the military was considered ineligible. The salary for Corpsmen was $100 per week and employment was only guaranteed for the duration of the federally-funded program.

The Fire Salvage Corps would be special-called by the officer in charge. A uniformed officer and chauffeur were assigned to each Corps unit. The officer would direct the Corpsmen at the scene of a fire. A total of five special vehicles outfitted with salvage tools and equipment were dedicated to the program. The major operational difference between the FDNY Salvage Corps and the New York Fire Patrol was that the Fire Patrol limited its salvage operations to commercial establishments while the Salvage Corps extended service to residential dwellings as well.

Corpsmen were provided with distinct dress and duty uniforms including a badge, plus fire scene protective equipment including a regulation helmet with a distinctive Fire Salvage Corps frontpiece. The badge was a simple two inch round disk with a burning house and the words "FIRE SALVAGE" in the center. Around the outer ring of the badge were the words "Fire Department - City of New York." An assigned number was stamped on the back of the badge near the hallmark. The badge has a pewter-like finish which, to date, has only been used for one other badge.

At least two other badges that may be mistaken as the official badge from the Model Cities program are known to exist. Both bear the burning house logo in the center. One is the standard FDNY Maltese cross and the other is a stock starburst pattern. Both of these badges were prototypes submitted to the Department for consideration. However, only the circular badge previously described was actually accepted, ordered, and issued by the Department to members of the Salvage Corps.

Federal funding for the Salvage Corps terminated in 1991. Even though the City attempted to continue the program, it was terminated the same year.

Another salvage program was instituted in 1976 when additional federal funds were used to rehire ninety-eight firemen previously laid-off during the City's financial crisis. They were assigned to twenty selected ladder companies and trained in salvage operations. By 1979, 235 firefighters were assigned in forty-seven ladder companies under the federally-funded program, with at least one qualified salvageman on duty for each tour at each of the designated companies. Since these personnel were regular members of the uniformed force, they performed standard firefighting duties and would redirect their efforts to salvage operations at the discretion of the officer in charge. No special badges or other identifying emblems were utilized. This program was discontinued along with other efforts of the FDNY to perform salvage functions.

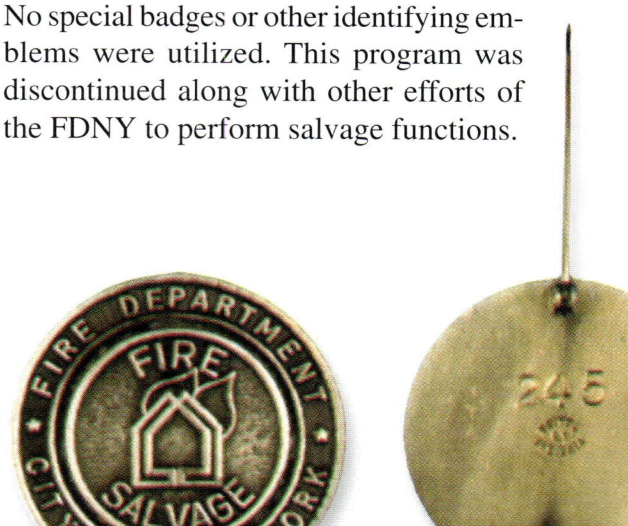

The Federally-funded Salvage Corps was made up of neighborhood civilians (1972-1991.) The badge number is engraved on back.

CHAPTER 12

Honorary Members

The designation as an honorary member of the New York Fire Department is coveted by many individuals. Although there are no official records of the earliest honorary members, we do know that some were awarded prior to 1850. In Costello's *Our Firemen*, mention is made of "Madame" Antoinette Otto being an honorary member of Engine Company 12 in 1850. Elmer Ellsworth, founder of the New York Fire Zouaves was an honorary member of Engine Company 14. A portrait shows an 1860 Engine Company 14 badge on his tunic. Similarly, photographs of his Adjutant, Major Charles McKnight Leoser, show him wearing an NYFD engine company badge, although he never lived in New York City until after his discharge from the army. Such honorary status was given by individual companies and it is not known whether this honor was also practiced by the Department. A badge with an uncertain origin (shown in Chapter 22) *may* be an early honorary member's badge from the 1860-1865 period.

1st Lieutenant (later Major) Charles McKnight Leoser of the 11th Regiment, NYS Volunteer Infantry, also known as the 1st New York Fire Zouaves. Although never a member of the NYFD, Lt. Leoser is seen here wearing an NYFD Engine Company badge of which he is believed to have been an Honorary Member. (Photo courtesy of the Gil Barrett Collection - U.S. Army Military History Institute.)

Custom, solid gold Honorary Deputy Chief Max Bamberger's badge, dated August 1927.

There is only one Honorary Commissioner and one Honorary Chief of Department at a time.

Appointment as an honorary member in the FDNY is bestowed upon individuals who have made a significant contribution to the goals of the Department. An additional criterion is that the individual must be a resident of the State of New York. Records could not be located to document the first Departmental appointment of an honorary member. A book about the FDNY with an unlikely title, *Firemen and their Pets*, states that Dr. Harry M. Archer was the first civilian to be appointed with an honorary rank of Battalion Chief in 1907, concomitant with the title of Honorary Medical Officer. In 1913, appointments as Honorary Chief were granted to William F. and Edward J. Kenny for donating two searchlight trucks and an ambulance to the Department.

The rank or title of the appointment is reflective of the individual's contribution or profession. There are numerous titles that may be issued with honorary status, including the following:

- All Commissioner titles
- All Chief ranks (except Deputy Assistant Chief)
- Medical Officer
- Chaplain
- Lieutenant
- Captain
- Photographer

Appointment to the two highest ranks, Chief of Department and Commissioner, may only be nominated by the Commissioner. The candidate must have been an honorary member for no less than twenty years. It is Departmental policy that there be only one Honorary Commissioner and one Honorary Chief of Department at a time.

A banker by profession, Honorary Battalion Chief Robert Mainzer was instrumental in the formation of Rescue Company 1 in 1915. For this, a testimonial dinner was given in his honor in 1916.

Becoming an honorary member of the FDNY is a serious matter. It is not taken lightly by the Department. Appointment includes a probationary period and a review once every five years to determine if status should be maintained. Promotions may be awarded depending on continued and additional services rendered to the Department. Appointments may also be revoked requiring the surrender of all Department-issued property including badge, identification card and certificate. An honorary member who maintains his or her status for ten continuous years is granted tenure, thereafter requiring evaluation only upon consideration for promotion.

While basic badge styles have been relatively consistent for honorary members, certain elements have changed.

The first known badge style for honorary officers was the pentagon. There were two differences between the badges for line and honorary officers. In the number plate the letters "HON" appeared for honorary officers. Also, many of these badges can be found with a solid back.

Honorary Battalion Chief pentagon (pre-1937.)

Early honorary officer badges with an "H" or "HON" engraved on a standard issue badge.

Badges with, "Honorary" cast into the enameled ring.

When the Department adopted the sun-ray Maltese cross with eagle badge for line officers, honorary officers continued to receive the pentagon badge for several years. However, some time in the early 1940's honorary appointees were issued the newer style badge. The lettering "HON" or simple "H" would appear in the number plate.

Variations of the badge include different treatment of the number plate. The "HON" has re-appeared several times in either enamel engraving or reverse enamel. At least one manufacturer enameled the entire plate in red. Others have left it blank or filled it with scroll work. The next change to occur was the addition of the word "Honorary" on the title ring in lieu of this designation in the number plate. This is the current style.

Custom, jeweler-made Honorary Deputy Commissioner badge.

Current badges for honorary officers with stars and leaves.

Several honorary officers have had badges made in the same starburst style used by some high-ranking officials of the Department.

Honorary Lieutenant and Captain - discontinued in 1995.

Honorary Lieutenant badge presented to Jess Strait of the New York Daily Mirror by Commissioner Quayle circa 1946.

Recently, stars and leaves have been added to the badges of several Honorary Chiefs and Commissioners. These badges have been made privately; they are not issued by the Department. Similarly, several current honorary officers including the Honorary Chief of Department and Honorary Deputy Commissioner have badges that are in the sunburst, or police style, that was adopted by the Department. The Department still has an inventory of Maltese cross with eagle style badges and it remains to be seen what, if any, future changes are adopted for honorary member badges.

Many physicians and clergy, who have volunteered their professional services to the FDNY have been appointed Honorary Medical Officer and Honorary Chaplain.

An unusual Honorary Lieutenant badge believed not to have been issued by the Department.

Early Honorary Medical Officer badge issued to Dr. Ernest Stillman adapted from 1937-vintage Captain badge.

The honorary ranks of Lieutenant and Captain were discontinued in 1995. For some undetermined time, the title of Honorary Dental Officer was also awarded, but is no longer.

The Honorary Fire Officers Association (HFOA) was founded in 1976. As of 1994, honorary officers are required to apply for and maintain membership in the HFOA as a condition of their appointment. The group continues to support and contribute to the FDNY just as each individual member has done. The organization is responsible for maintaining all administrative records pertaining to honorary members of the FDNY. A select committee from the HFOA acts as advisors to the Commissioner regarding the appointment of honorary officers and the promotion of existing honoraries.

Monsignor John Reilly being appointed as an Honorary Chaplain by Commissioner Cavanaugh and Chief of Department Loftus.

Many freelance photographers develop a reputation and portfolio of fire photography. Not surprisingly, many of them are Buffs. Based on their contributions to the Department, these photographers may be given the title, Honorary Photographer. This is the only non-officer honorary title that is bestowed by the Department.

Badge of an Honorary Photographer.

Badge of Harold V. Smith, President of Home Insurance Company.

One of the more active and visible honorary officers of the Department's history was Honorary Assistant Chief Clarence Meek. Chief Meek functioned as the Department Librarian/Historian from 1938 until his death in 1973. Until recently, the Department library was named and dedicated to Chief Meek. Chief Meek was initially appointed at the rank of Honorary Captain in 1938. In 1953, he was promoted to Honorary Battalion Chief, followed by Honorary Deputy Chief in 1959 and finally to Honorary Assistant Chief in 1967. Many of the articles written by Chief Meek served as valuable historic resources for the material contained in this book. A Fire Lines badge belonging to Chief Meek is included in Chapter 13.

The badges of three well-known honorary officers are shown. The first belonged to Harold V. Smith, President of Home Insurance Company and the New York Board of Fire Underwriters. As noted in Chapter 6, his prominent role in the fire insurance industry earned him a place in the history of the fire service. Harold Smith's love of the fire service led to his unparalleled collection of memorabilia found in several fire museums, including the New York City Fire Museum and the American Museum of Fire Fighting. Many of the badges shown in this book come from his collection. Mr. Smith's Honorary Deputy Chief badge has a flat back inscribed with his name and the address of the Home Insurance Company, 60 Maiden Lane, where Mr. Smith also maintained a residence.

The second belonged to Kenneth Straus, Chairman of the R. H. Macy Company. Macy's is New York's first "department store." Its

Ken Strauss receiving his badge as Deputy Commissioner from Commissioner Edward Cavanaugh as Chief of Department Arthur Mallet looks on.

flagship store in Herald Square is still the largest department store in the world. The company was founded in 1858 by Rowland H. Macy and acquired in 1888 by Isidor and Nathan Straus, Kenneth's great-grandfather and granduncle. Isidor Strauss lost his life on the *S.S. Titanic*. Kenneth Strauss was an avid FDNY Buff and donated untold dollars in support of its members. He also gave his personal time to many social and philanthropic activities related to the Department. For this, he was awarded the rank of Honorary Deputy Commissioner and later, Honorary Commissioner. A portion of his private collection of fire memorabilia is on permanent display at FDNY headquarters.

The third badge belonged to Edward Waterman. Mr. Waterman was a professional photographer employed by the New York City Board of Estimate. This job took him all around the City. Being a Buff, he frequently photographed fires and fire department subjects. His photos have been frequently published in books on the FDNY and on fire service history. Mr. Waterman was a Buff of Engine Company 219 in Queens. His unusual badge has a large plate added on top of the number plate bearing an inscription in red enamel of the full word, "Honorary."

Honorary Medical Officer Harry M. Archer in his unique white uniform receiving the Department's highest honor, the James Gordon Bennett Medal, for rescuing two workers at a tenement collapse on Eldridge Street in 1923. It was his third medal. (Photo courtesy of FDNY/Fire Safety Education Fund.)

Honorary Battalion Chief badge of Ed Waterman, famed fire photographer and FDNY buff

For many years, honorary officers were required, upon appointment, to purchase a dress uniform with accouterments commensurate with their honorary rank. And, they were required to wear them when ordered. Special Order #159 dated September 1953, orders "Honorary Deputy Chiefs In Uniform" to be included in the funeral procession for Assistant Chief of Department in Command John McCarthy. A 1912 photograph of Dr. Archer shows him in a white uniform. Dr. Archer had this uniform custom-made and it is the only known white uniform to have been worn by any honorary or regular member of the Department. Honorary officers are no longer required to possess a uniform, although they are permitted to have them, and several do.

It was also once standard practice for Honorary Chiefs to be issued a Maltese cross car badge. Early NYPD Procedure Manuals include these in the section on "Automobile Identification Plates" (see Appendix). These car badges were identical to those issued to Department-owned vehicles but have since been changed as described in Chapter 17. The privilege of displaying car badges has been greatly curtailed over the years and is now limited to high-ranking honorary officers as approved by the Fire Commissioner.

Community Relations badges issued in the 1960's to selected individuals.

For a short period of time, around 1982 to 1987, badges were made with the words, "Community Relations" in the title ring and "HON" in the number plate. These were given to members of the community who assisted the Department in some respect but who did not qualify for the more stringent honorary member status. Two variations of this badge were made, one with standard FDNY red enamel and one with green enamel. There appears to be no significance to the two colors. The Community Relations Bureau was originally created in 1966 by Commissioner Robert O. Lowery.

Honorary Deputy Chief Gordon Mullins and his red 1950 Nash Ambassador with flashing lights, siren and vehicle badge #13. (Photo courtesy of The Gordon Archive.)

A very unusual badge that is probably not Department-issued.

Badge of Gordon Mullins with Public Relations.

Two Community Relations badges that were located are very unique but their authenticity cannot be confirmed. They are believed to have been made privately. The first is a very attractive badge in gold. It is a sun-ray Maltese cross design but has a standard FDNY Maltese cross in relief in the center. The wording on the badge is not consistent with Departmental badges.

The second bears a person's name in the upper panel, Gordon R. Mullins. Mr. Mullins was a life-long Buff, and Honorary Deputy Chief and eventually an FDNY Dispatcher in the 1960's. He is known to have had many different badges over the course of his life.

On rare occasions, honorary firefighter badges have been issued. They are not issued by the Department. Rather, individual companies have elected to present these badges on their own, as they did in the 19th Century. Typically, honorary firefighter badges are issued to youngsters. Recently, a badge resembling a standard Firefighter's Maltese cross with the word "Honorary" engraved in the central disk was awarded by Ladder Company 176 to a disabled youth in the community who was very fond of the company and they of him. Two examples of these unauthorized badges are shown in Chapter 18.

Tonight Show *host Johnny Carson receiving an Honorary Chief helmet from Commissioner Robert O. Lowery in 1967.*

CHAPTER 13

Fire Lines Badges

Several badges were authorized by the Department to admit non-Department personnel to the scene of fires. In the early days of the Department, after the introduction of badges for firemen, badges were only issued to individuals who had a legitimate, occupational need to get close to a fire scene. Basically, these were officers of insurance companies and press reporters. However, since the turn of the century, badges have been issued as status symbols to persons with political connections in the City.

The earliest badge described for, "persons to be admitted at fires other than members of the force," is described in the 1865 "Act to Create a Metropolitan Fire District and Establish a Fire Department Therein." It was issued to "the Fire Marshal and Assistant Fire Marshal, Superintendent and Inspectors of Unsafe Buildings, Inspectors of Gas, reporters for the press and the officers of each insurance company" as described below:

> *The badge is of yellow metal, in the form of a shield, surmounted with fire cap and crossed pipes. In the centre of the badge, within a shield, is a design representing the seal of the Department. Over this, in raised letters, are the words, "Metropolitan Fire Department." Beneath are large figures in white metal, designating the number of the badge.*

The 1868 Annual Report of the Metropolitan Fire Department describes these badges as fire lines badges and states that 210 such badges were issued by 1868. It is not known what the final count was by the end of the State's control of the Department in 1870. The Report reiterates the description of the badges as first appeared in 1865. An interesting and unique feature of this badge is the shield representing the seal of the Department which incorporated both the New York State and New York City seals illustrating the State's control over this City agency.

An example of another version of a fire lines badge specifically for insurance company executives is shown in Chapter 6. The badges are significantly different from

Badge issued in 1865 to "persons who are not members of the force" for admittance to fire lines.

the description in 1865 and 1868. They are a simple round silver badge with the name of the insurance company and the officer's title. According to Costello, this same badge was issued to press reporters but was engraved with the word "Reporter" and the name of the newspaper.

1884 Department-issued Fire Lines badge.

Apparently, there was lax control over the issuance of fire lines badges, or certain individuals took it upon themselves to have badges made without the approval of the Department. On May 20, 1884, General Order No. 4 of the Board of Fire Commissioners, required the replacement of all previously issued fire lines badges with one uniform badge. The text of the Order is very interesting as it tells a lot about the badge and its use. The full text of the Order is included in the Appendix.

The Order indicates who may be issued fire lines badges:

> **IV.** The badges are issued upon payment of a deposit of two dollars for each one, as follows:
>
> To each local fire insurance company not more than three.
>
> To each principal agency in this city of fire insurance companies located outside of the city, whether it be an agency of one or more companies, not more than two.
>
> To each newspaper published in the city, or press association located therein, such a number as the Board may determine.
>
> To such public officials and others as the Board may direct.
>
> **V.** The badges will also be issued, without deposit, to the officers of the uniformed force; and to other subordinates of the Department not in the uniformed force, without deposit, upon their application granted by the Board; in both cases on condition that the penalty for losing the badge shall be a fine of five dollars.

The "entertainment" aspect of fires is documented in this 1905 postcard entitled, "Crowd at a Fire, New York."

A group of civilians watching a fire being fought by men of Engine Company 144 (Brooklyn) circa 1911. The civilian on the left is wearing an NYFD Fire Lines badge. (Photo courtesy of Steven Scher.)

While the Order seems clear about who can be issued a fire lines badge, there is an obvious loophole allowing "others as the Board may direct." This caveat allowed the Board and other city officials to issue a fire lines badge to their friends. Being authorized to cross police lines at the scene of a fire was clearly a sign of power, importance, and political connections. For early Buffs with political connections, a fire lines badge was a coveted item.

Five standard fire lines badges were issued by the FDNY. They are commonly referred to as the 1884, flame-top, police-style, star-in-circle and alarm box. Of these, the date of issuance of three are definitive. The first documented fire lines badge was issued May 20, 1884. The police style badge was issued on March 15, 1900 and the star-in-circle was issued on January 1, 1907. Copies of documents relative to the 1884, star-in-circle and police-style fire lines badges are included in the Appendix.

Many collectors believe that the flame-top preceded the 1884 badge. Information about the badges proves otherwise. One badge shown was issued to the New York newspaper, "Morning Advertiser." This newspaper was only published from 1891 to 1897, dating the badge to this period. The fire lines badge issued to Clarence Meek is of the flame-top design. Honorary Chief Meek was probably issued this badge in the early 20th Century during his tenure with the Ahrens-Fox and/or the Mack fire truck

Alarm box Fire Lines badge circa 1905-1915.

1900 pinched-shield Fire Lines badge.

Star-in-circle badge issued in 1907.

Fire Lines badge issued 1892-1897 to the "Morning Advertiser" newspaper.

"Flame-top" Fire Lines badge issued to individuals. Badge on left belonged to Clarence F. Meek, later an honorary officer and Department Librarian and Historian.

companies. He was appointed an Honorary Captain and began serving as the Department's librarian in 1938. Therefore, it is likely that the flame-top style, as well as the tombstone/alarm box and star-in-circle were used from the late 1800's well into the teens of the 20th century. However, there is no hard data to confirm this assumption.

An interesting feature of the flame-top badge is the use of two numbers on these badges. The first, in a disk near the base of the badge, represents the Department-issued badge number. For badges given to companies, (e.g. insurance companies and newspapers) this number represents the company. A second number in the disk under the banner with the company name is the number of the badge *within* the company. Badges issued to individuals have their name in the banner, the Department-issued number in the lower disk and the number "1" in the upper disk.

BROOKLYN DAILY EAGLE

APRIL 1, 1899
Badges For Retired Firemen

Commissioner Scannel Says They May Be Very Useful at Fires

Fire Commissioner Scannel has decided to permit firemen who have served twenty years or more in the Fire Department and who have been placed on the retired rolls to obtain fire lines badges. At present a fireman who has had many years of experience, in fire fighting, and who lays aside his uniform, is immediately barred from appearing at fires where his knowledge might be of assistance to citizens and to the department.

Commissioner Scannell's action is due, in a measure, to the experience of a number of retired firemen at the Hotel Windsor fire. They were among the first on the scene and while instructing citizens how to hold life nets and to use life-saving apparatus they were hustled away by the police because they had no badges on.

To-day was pay day for the retired men at Fire Headquarters and when the veterans learned of the action of the Commissioner they were greatly pleased. They stated that it would be their aim to give every assistance at any fire to which they could get.

Commissioner Scannell states that he considers that any officer or fireman, after faithfully serving the city for more than twenty years, is entitled to such a privilege and that men of such experience in fire fighting would be of great assistance on such occasions as the Windsor Hotel catastrophe. The retired men will make application for the badges in the regular way and the interest on the $5 deposit in each case will go to their pension fund.

Within two weeks of the March 17, 1899 Hotel Windsor fire, Commissioner Scannel, impressed by the efforts of the retired firemen at the blaze, initiated the issuance of fire lines badges to those who had served twenty or more years of active duty. This fire is a significant episode in FDNY history due to numerous unusual circumstances related to the blaze, including the participation of off-duty and retired firemen. A small fire in a sitting room had time to spread unnoticed because the majority of the buildings occupants were watching the St. Patrick's Day parade pass below along Fifth Avenue. The parade itself hampered the response of the Department and the high occupancy of the building led to a number of dramatic rescues as well as forty-five deaths and fifty-two injuries.

The dramatic and deadly fire at the Hotel Windsor prompted Commissioner Scannel to offer Fire Lines badges to retired members of the FDNY.

Unusual and rare celluloid Fire Lines badges circa 1906.

It is not certain what fire lines badges for retired firemen looked like. They may have been standard civilian pentagon badges. However, in lieu of a title, the upper banner was engraved with the fireman's name and a badge number appeared in the number tab above. One such badge was found but how it was used cannot be confirmed. It is not known if the practice of issuing badges to retired firemen was popular or how long it lasted. There are many badges found with the designation "retired" on them, but they are not issued by the Department.

In his 1916 book, *The Boys Book of Firemen*, Irving Crump describes a group of rather wealthy and well-connected Buffs who possessed fire lines badges, "*similar to those used by members of the press.*" It describes these Buffs as having an elite club. What is most interesting is that this reference pre-dates the establishment of the first authorized and recognized Buff organization, the Fire Bell Club of New York, formed in 1939 (see Chapter 16).

For a short period, non-metallic fire lines badges were issued. These were made of a laminated pressed paper. Two examples of these non-metallic badges are shown, one with the silhouette of a Maltese cross and one a simple oval shape.

This is believed to be the Fire Lines badge issued to retired firefighters in 1899 by Commissioner Scannel.

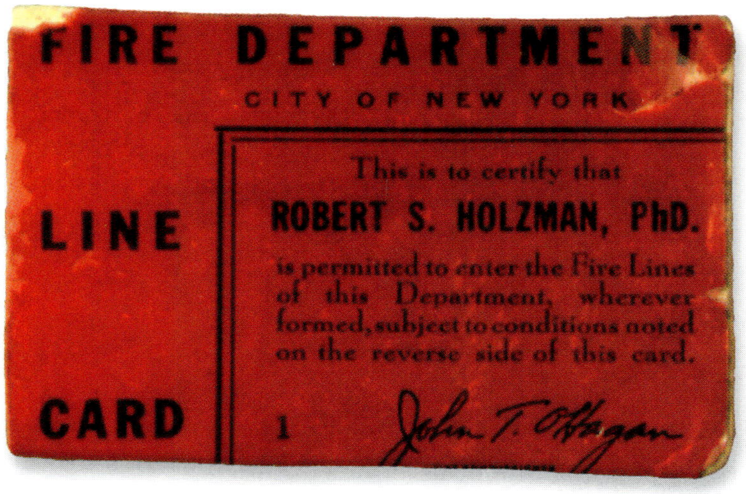

Fire Lines card #1 issued to Dr. Robert Holzman, author of "The Romance of Firefighting."

The practice of issuing fire lines badges ended around 1915. However, the practice of admitting selected individuals to fire lines was accomplished with a printed card in place of a badge. Conditions for the use of the card are specified on its back:

> "A person receiving this fire lines pass in accepting and using it, expressly agrees that:
>
> 1. He assumes all risk of accident.
> 2. It is for the exclusive use of such person and will be forfeited if presented by any other person and may be taken up at any time for cause.
> 3. Said person shall not use any facsimile or other copy of said pass on any vehicle operated by him.
> 4. The holder of this pass shall not at any time, while within fire lines, be permitted to enter fire or adjacent buildings."

In addition, uniformed personnel and city officials were, and continue to be, allowed within fire lines. Individuals with authorized Press credentials issued by the New York City Police Department may be admitted at the discretion of the police official in charge at the scene.

Several unique fire lines badges exist. One totally custom badge is believed to have been issued by a fire alarm company some time around the 1898 consolidation of the City. Within the center disk are the very fancy, intertwined letters, "GNY" for Greater New York, the name by which the City came to be known upon consolidation. The wording on the badge which permits the bearer into buildings is also unique. The badge bears the name *Fire Protective Co.* Badges from other fire alarm companies are included in Chapter 17.

Another interesting fire lines badge belonged to Ed Waterman, famous fire photographer and FDNY Buff. He possessed a badge made from the standard FDNY pentagon with the words "*Admit to Fire Lines - E. A. Waterman*" inscribed in the rank/title banner. Mr. Waterman had many friends within the Department and was a significant recorder of FDNY history.

It is unusual to find fire lines badges from non-FDNY departments in NYC. The only definitive badge located is one from the Long Island City Fire Department. Long Island City existed as a paid department from 1891 to 1898, which is the presumed date range of this badge. The example does not have an individual or company inscribed in the panel and it bears the number "1." It is questionable whether it, or any other fire lines badge was ever issued by the Long Island City Fire Department.

A Brooklyn Fire Department equilateral cross with the word "Special" inscribed may be a fire lines badge. However, no documentation confirms the precise use of this badge. It is shown in Chapter 4.

Unusual "Admit to Buildings" Fire Lines badge for a private fire alarm company.

Special Fire Lines badge allowing admittance to buildings given to Edward Waterman, famed New York fire photographer and official photographer for the NYC Board of Estimate.

Rare, unissued Fire Lines badge from the Long Island City Fire Department.

CHAPTER 14

Buffs

"Buff (buf) n. [< It. bufalo, buffalo] 1. a heavy, soft, brownish-yellow leather 2. a military coat made of this. 3. a dull brownish yellow. 4. [Colloq.] a devotee; fan" from Webster's New World Dictionary of the English Language.

While the history of fire Buffs has not been documented, it seems safe to assume that they have existed in the United States, in one form or other, since colonial times. Today, it is difficult to define the typical Buff. While it has generally meant someone who is dedicated to, but not a member of the fire service, this may no longer be the case. Many firefighters have an interest beyond their assigned duties, and many have become actively involved in collecting fire service memorabilia. To that degree, they too are Buffs.

The earliest reference to Buffs, at least by description if not by name, appears to date back to the days of hand-drawn apparatus in New York. During this era, "runners" hovered around firehouses and upon an alarm ran with the company, helping to pull the apparatus. Unfortunately, the runners had a reputation for rowdyism and looting. They became such a hindrance to the Department, that on June 21, 1824, a resolution was adopted by the Common Council for the Chief Engineer to advise all companies to disperse such

Oil painting of Chief Engineer Harry Howard (1857) wearing a buffalo coat purported to be the source of the word Buff. (Photo courtesy of the NYC Fire Museum)

"volunteer associations" and to no longer make use of their services. This was reiterated in ordinances in 1839 and 1841. One section of the ordinance states:

> *"Resolved, That the practice of permitting volunteer associations to assume the garb of firemen, and to mingle in the duties thereof, is not only in direct and open violation of the ordinances of the City Council, but calculated, in its results, to demoralize the character of our youth, and bring reproach upon the Department, by the riotous and disorderly conduct in which they are so often engaged, and that the officers and members of each company be desired forthwith to disband all associations of volunteers, and upon no occasion to suffer or permit them to have access to the public property; and all magistrates, watchmen, and police officers are hereby requested to prevent the congregating of all boys around, or in the vicinity of engine, hose, and hook and ladder houses, to the end that members of the Fire Department may be recognized as such, and be held responsible for all deviations from the path of duty, and the requirements of the ordinances of the City Council."*

For the most part, the relationship between the volunteers and the companies to which they were attached was symbiotic. However, the rowdy element overshadowed the good. As time went on, these "boys" became a mounting problem for the Department. In 1842, the ordinance was again amended and reinforced:

> *"No boys or other persons known as volunteers, shall be permitted to wear the garb of firemen, have access to any of the places of deposit of fire engines, hose, hooks and ladders, or other apparatus of the Fire Department, or run as members with any such engine, hose cart, or hook and ladder truck, or form any associations similar in any way to those now known as "Volunteer Associations," and any fire company in the city of New York which shall consent to any violation of the provisions of this section, shall be forthwith disbanded."*

Early Buffs continued to have a reputation as nuisances, even after the "Volunteer Associations" were prohibited, as attested to by Chief Engineer A. G. Carson in this 1854 report to the Common Council:

> *"Rowdies still continue to attack the firemen and the latter occasionally attack each other. Companies should be held responsible for the villainy of those whom they permit to run with their engines and to congregate in and around their houses."*

Young boy soon to be a runner or buff visiting the local fire house and listening to stories told by the firemen.

However, given the restriction of age and limitation on company size, many young men who could not become firemen fulfilled the role of runner. In fact, many famous New York firemen, including Chief Engineer Harry Howard, were runners before becoming regular members of the Department.

New York was not the only place where runners, or Buffs, created a problem. The City of Brooklyn had similar problems. An Ordinance to Establish a Fire Department in the City of Brooklyn, dated November 15, 1869, stated:

> *"It shall be the duty of each and every fireman, to prevent boys or disorderly characters from congregating in or about the place of deposit of their various apparatus, and no person will be allowed hereafter to meet or assemble in any of the houses connected with this department, other than members and exempt members of the department, or persons over eighteen years of age introduced by the members of the company so visited, who shall be responsible for their conduct during such visitation. It shall also be the duty of all firemen to see that good order is preserved in and about the houses occupied by their respective companies, and to prevent any person from habitually congregating on walks in front of the houses or their said apparatus.*
>
> *No fireman or other persons shall ride on any of the apparatus, either in going to or returning from a fire, or alarm of fire, or upon any other occasion. The drag-rope shall be the proper place for the firemen, except officers in command, and it shall be their duty to prevent all boys and noisy and improper persons from taking hold of the rope or apparatus. And no person other than a member or exempt member of the department shall be allowed to take the tongue or tiller, or have any command of any apparatus."*

Similar accounts are found in writings of the Morrisania Fire Department in the Bronx, circa 1865.

From J. F. Kiernan's, *Reminiscences of the Olde Fire Laddies of New York and Brooklyn* (1885) comes the following passage:

> *"The bane of the Volunteer Department of Brooklyn for many years was the system of runners. They used to love to hang around the engine-houses, and upon the slightest opportunity would fall into the ranks with the regular firemen. Their principal joy was to run with the engine. They assisted in pulling the engines and originated nearly all the fights. They, having no responsibility to answer for, and no character to lose, thought it a good thing to spoil for a fight on all occasions, and finally became so bad that two fire companies could not meet each other without a fight occurring. The runners were the first in it and always the first out of it, leaving the regular firemen at loggerheads, and of course it was upon those that all blame rested. In consideration of all this, it was forbidden for a company to allow any but its own members to run the engine, and all outsiders were prohibited to pull the rope."*

Although this appears to be the history of the Buff, the actual word is never found in historic documents of the time or books of the era. The exact timing and introduction of the term Buff is unknown but perhaps the earliest printed reference to the term appears in the book, *Firefighters and their Pets,* published in 1907 by Alfred Downes.

The word Buff is believed by many to have originated in the fire service. Even if this is true, how it came to be is a subject of much debate. The most prevalent legend links this name to a type of coat worn by firefighters of a century ago, as described in the dictionary definition. The 1839, 1841 and 1842 ordinances specifically mention the wearing of firemen's "garb" which may well have meant their buffalo coats. Numerous dictionaries make the connection between a devotee of the fire service and the buffalo, or buff, coats. A similar connection is made in reference to runners in A.E. Costello's book, *Our Firemen.* In it, he reports that the engine of Humane Engine Company 20 was retired to the Corporation Yard as penalty for permitting "the boys in the regular uniform to run with them."

The New Oxford Shorter Dictionary actually attributes the origin of the word Buff to the NYFD:

> *"8. Orig., an enthusiast for watching fires (so called from the buff uniforms of volunteer firemen in New York City)."*

Whether or not the origin of the word derives from the coats is debatable since no historic references or writings from the period specifically make this connection. It is interesting to note that two other major dictionaries give the same definition as the Oxford dictionary, but disagree as to the time of origin. Websters New International Dictionary (Unabridged) indicates the period for wearing buffalo coats as circa 1820. The American Heritage Dictionary of the English Language dates it as 1920. This disparity raises a reasonable doubt about this theory of the word's origin.

Another possible connection between the words Buff and buffalo comes from the autobiography of Alfred E. Smith, one-time governor of New York, 1928 Presidential candidate, afficionado of the fire service throughout his life and a Buff of Engine Co. 32. His father and uncle were members of Liberty Hose 10 of the old volunteer department. In his autobiography he writes:

> *"After the great blizzard of 1888 the schools were closed and there was not much for the boys to do, so I became a volunteer worker in No. 32*

Engine Company, located on John Street. Volunteer work of that nature was agreeable to me because I had always had a strong desire to be a fireman and I performed my duties under what was then called the Buffalo Corps. The Buffaloes was an order of young men interested in fire fighting and in the glamour and excitement of the engine house."

Although Gov. Smith does not specifically use the name Buff to refer to the young men of the Buffalo Corps, the late Honorary Deputy Chief Clarence Meek, the recognized authority of FDNY history, shared his belief about the origination of the term. Gov. Smith's account makes the clear connection between the "Buffaloes" and the earlier volunteer organizations. The date of the formation of the Buffalo Cadets is not known but, if it was in 1888, it may explain why the term Buff is not found in books published before that year.

Dr. Robert Holzman, author of *The Romance of Firefighting* suggests a variation on this theory. He notes that it was common for members of the Volunteer Fire Department to wear the buffalo coats previously described but that the paid Fire Department adopted the now common blue serge uniforms. Volunteers who did not become members of the paid Department but who continued to be permanent fixtures in the firehouses, came to be known as Buffs because of the coats that differentiated them from the uniformed force. Dr. Holzman was a Buff of Engine 56 in Manhattan for many years, the same firehouse that the legendary Dr. Harry M. Archer called home. Dr. Holzman's father was also a Buff back in the time before the consolidation of the City and the New York, Brooklyn and Queens fire departments.

John V. Morris, noted fire historian, reported yet another possible origin of the term Buff in his book, *Fires and Firefighters*. Morris attributes the original use of the term, like Holzman, to the transition period from volunteer to paid firemen in New York City. However, he believes it was used to refer to volunteers who continued to inhabit firehouses after the conversion to the paid force and who spent hours "buffing" their cherished "masheens." Paid members, Morris theorizes, referred to these volunteers as the "Buffs."

During World War I, Buffs were actively recruited to become members of the Fire Auxiliary Corps. Under the Department's official classification system for recruiting, Buffs were described as follows:

"Fire Buffs. Men who out of high interest and motive for the service have gained a fine degree of knowledge and technique from their attendance at fires and who are known as "Fire Buffs." Fire Buffs need no further appeal than this. The gallantry and self sacrificing spirit which has characterized them in their relation to the "Fire Game" is still alive. These men are requested to volunteer their services stating hours they can give, when and location in which they prefer to serve."

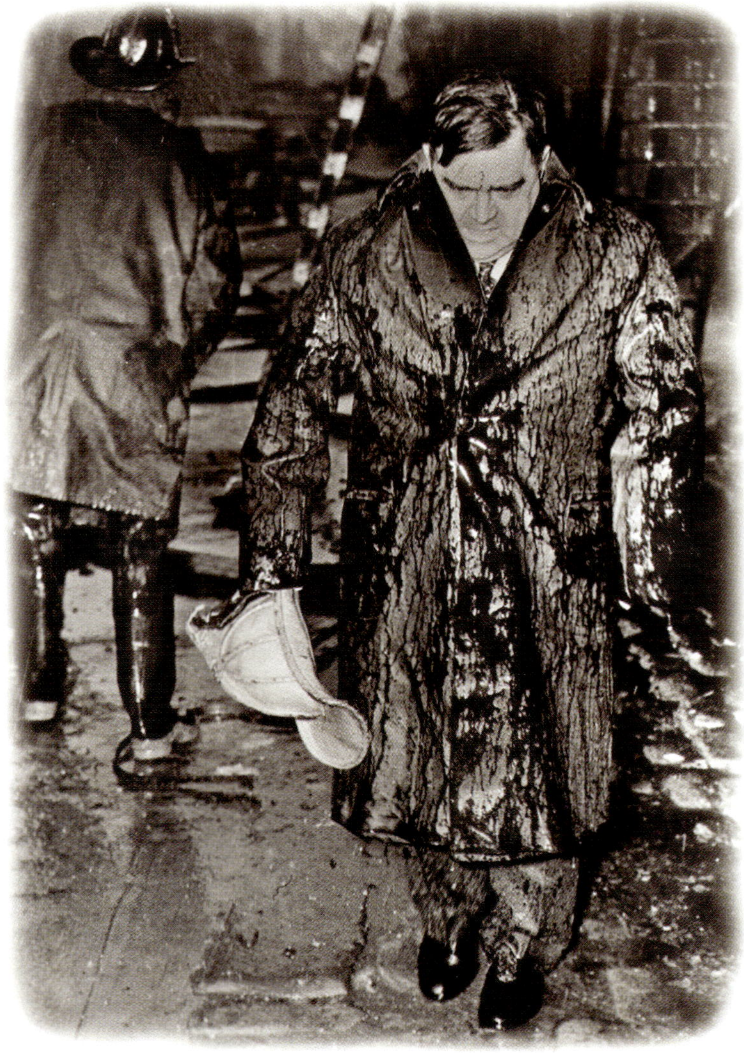

Perhaps the most ardent and well-known FDNY Buff, Mayor Fiorello LaGuardia, in 1939. (Photo courtesy of Brown Brothers.)

A rather curious comment regarding Buffs was made by Special Deputy Commissioner Eli Joseph in an August 1918 letter to commanding officers and members of the Fire Auxiliary Corps. In it he states, *"The day of the Buff is gone, and your co-operation and aid is earnestly asked for and generously appreciated."* This statement can be taken to mean one of two things: that the rowdy days of the late 1800's would not be tolerated but earnest cooperation was invited; or that the days of simply hanging around the firehouses were over and volunteers would be pressed into active duty. Given that the official classification system holds Buffs in high regard, it probably reflects the Commissioner's intent. Either way, Buffs have made their mark on the Department and its history.

One of the most visible Buffs in the Department's history was New York City Mayor Fiorello LaGuardia. During his tenure, the Mayor's interest in the Department far exceeded that called for by his duties. He often raced to greater alarm fires. During the war years, as Director of

Civilian Defense, he sent three FDNY members to London to observe firefighting under war conditions. Upon receipt of their report, he issued a handbook to the Mayors of other cities outlining the appropriate preparations in the event of fires resulting from enemy attacks. The Fire Auxiliary Corps (see Chapter 11), developed under his orders, was one of the largest and best-trained in the country, rivaled only by the Auxiliary Fire Service of England. Many of the auxiliarists were or later became Buffs and formed some of the Buff organizations of the post-war period (see Chapter 16).

Regarding FDNY Buffs and badges, there is no historical documentation on when Buffs began possessing badges. However, most hard-core, traditional Buffs, those who routinely inhabit a dedicated firehouse as the company's adopted mascot, possess a badge. Although a Buff may independently have one made, it is common, though a violation of regulations, for a Buff to be presented a badge by the members of the company.

Section 13.1.19(M) of the *Rules and Regulations of the Uniformed Force of the Fire Department of the City of New York* circa 1955 states:

> "Members shall not make a presentation of a buff badge to any person."

One hundred years after being designated *persona non grata*, the Department has acknowledged the existence of Buffs but still strongly objects to any practice that encourages or endorses Buffs who mimic or represent themselves as being closely or officially affiliated with the Department. The current Rules and Regulations still prohibit the issuance of any badge by a member of the Department to anyone, but it no longer specifically uses the term Buff.

Given their long standing and very specific prohibition, there is no standard design for Buff badges. They typically include the company number, the word Buff, and/or FDNY, in the top and bottom arm of the cross, or the center disk.

Generic FDNY Buff badges.

Badge personalized with the Buff's initials.

Company-specific Buff badges.

Badge presented by Ladder 47 to their buff "A. Guerrero" The "C" catch suggests this badge is from the early 1900's.

This is the epitome of FDNY Buff badges, a standard FDNY firefighter's Maltese cross with the word, "BUFF" in place of a badge number.

However, some Buffs have been successful in purchasing a badge that is virtually identical to the official FDNY Firefighter's Maltese cross but with "BUFF" in die-cut lettering in lieu of numbers. All of the FDNY Buff badges found while researching this book were in the shape of either a Maltese cross or a pentagon.

Very unusual and unique badge identifying a "Lady Buff" from Engine 276.

Perhaps the most intriguing of all the Buff badges found is one in the collection of the NYC Fire Museum. The upper arm of the badge is engraved "Lady Buff" and the lower arm displays "FDNY." The number "276" is applied in the center. Extensive research failed to uncover the original owner of this badge.

Edward Waterman's photographs of the FDNY are historic. He was employed by the NYC Board of Estimate as a photographer but his profession also carried into his hobby, buffing. He was an active Buff of Engine 219 which was the subject of many of his pictures. Living around the block from the firehouse was convenient and enhanced all the more with the help of some friends in the Department who installed a bell circuit to his house. Mr. Waterman was also an active volunteer in Rockaway Beach Protection Hook, Ladder, and Hose Company 1. His Buff badge was not a standard size but rather a small or "courtesy" size which is shown here.

The oldest Buff badge located belonged to Joseph F. Mastrella who buffed Engine 11 until about 1915. His son, Joseph F. Mastrella, Jr. joined the Department and went on to become a Battalion Chief, retiring in 1987.

This is believed to be one of the oldest Buff badges, circa 1915.

Buff badge of Ed Waterman.

Famous photograph of the fire horses of Engine Co. 219, taken by the company's buff, Ed Waterman.

The author presenting a first aid kit to "his" company Ladder 156, in 1970.

Several pentagon-shaped Buff badges exist. However, those found have the rather unusual designation of "Honorary Buff" - a rather redundant title. The two badges show here have names inscribed on the reverse. One is dated 1931 and states that it was presented by the NYFD Officer's Association. The origin of these badges is unknown and their authenticity has been questioned by some collectors and historians.

Two unique and unusual Buff badges in the pentagon shape used by the FDNY from 1884 through the 1940's. The badge on the left is inscribed, "Presented to Mr. John A. Tracy by Officers Ass'n. N.Y.F.D. June 25, 1931." The badge on the right is inscribed with the name, "Charles Burgess."

Two badges shown here do not bear any clear designation that the owner was an FDNY Buff. However, in both cases, the owners are known. The badge from Ladder 156 is the author's own badge, given to him in 1966. The second, which displays a blank panel in the top arm and only the words "FIRE DEPT" in the lower, was owned by the grandfather of the current Superintendent of the New York Fire Patrol.

There are, and have been, several associations organized by and for Buffs of the FDNY in the 20th century. Most of these issue a badge as a symbol of membership. In 1940, Commissioner Frank Quayle officially authorized only one such organization to issue badges, The Fire Bell Club of New York. Details about the badges of Buff organizations is included in Chapter 16.

Possession of a badge gives a Buff recognition, pride, and a sense of belonging. Since the typical Buff is someone who has strong admiration for the fire service and the company with which he or she identifies, the badge is a cherished possession and a symbolic connection to this avocation.

Unfortunately, the real origin of the term Buff may have been embellished or forgotten over the years and it may be less interesting than the Buff legends. After all, the definition of the word simply means a devotee or fan of any fancy. According to the Oxford Dictionary of the English Language, the oldest recorded use of the word Buff dates back to 1708. It was simply a colloquialism for a "dronish fellow." Although much less colorful and with absolutely no connection to the fire service, this very basic definition of a Buff may have been the case all along.

The author's Buff badge.

Although it bears no FDNY designation, this badge belonged to the grandfather of the current Superintendent of the New York Fire Patrol.

CHAPTER 15

Line Organizations

Line organizations are fraternal, religious and labor groups sanctioned by the Department. Line organizations are formed by members of the fire department who share some commonality. Five organizations representing union personnel include firefighters, officers, Fire Marshals, fire alarm dispatchers and supervisors and Marine Engineers and Pilots. The other groups, religious and fraternal, have more social purposes.

Line organizations do not routinely issue badges as a symbol of membership. However, many have adapted one of the two FDNY badge shapes for their logo. Logos may be reproduced in metal to look like an issued badge, and may be used as awards, gifts, or presentation pieces.

Some badges that have been produced by line organizations are shown in this chapter. In some cases, they are the size and style of a lapel pin only. The logo of the organizations that use a stylized badge are also depicted. Much of the information in this Chapter is taken from an article on FDNY line organizations which appeared in the second 1978 issue of *WNYF*.

Uniformed Firefighters Association

The Uniformed Firefighters Association (UFA) has been the labor organization representing line firefighters since March 1917. Only uniformed personnel with the rank of firefighter can become members. As a labor organization it is concerned with the health and welfare of its members and acts as the bargaining unit in contract negotiations with the City. It was the UFA that coined the term, "New York City's Bravest" to describe this city's firefighters.

The logo of the UFA is the standard FDNY firefighter's Maltese cross with the letters UFA in the central disk. UFA badges can be found in many versions, in varying sizes, metals, and configurations. They have been issued for a number of purposes, including gifts.

Assorted UFA courtesy badges.

UFA conference ribbon with badge.

Lapel badge of UFA Ladies Auxiliary.

Albert E. Guinness, first President of the UFA.

Badges of UFA elected officers.

Uniformed Fire Officers Association

Similar to the UFA, the Uniformed Fire Officers Association (UFOA) represents the officer ranks of the Department as its labor organization. It was founded on July 1, 1944. Prior to the founding of the UFOA officers belonged to individual rank-specific organizations like the Fire Lieutenant's Association and the Fire Captain's Association, which were organized near the turn of the century. (The Fire Lieutenant's Association was chartered in December 1918 as Local 148 of the International Association of Fire Fighters.) The logo of the association is an FDNY officer's badge with the lettering "UFO ASSN" in the central disk and the motto, "In Unity There is Strength

Lapel and courtesy badges of the Uniformed Fire Officers Association.

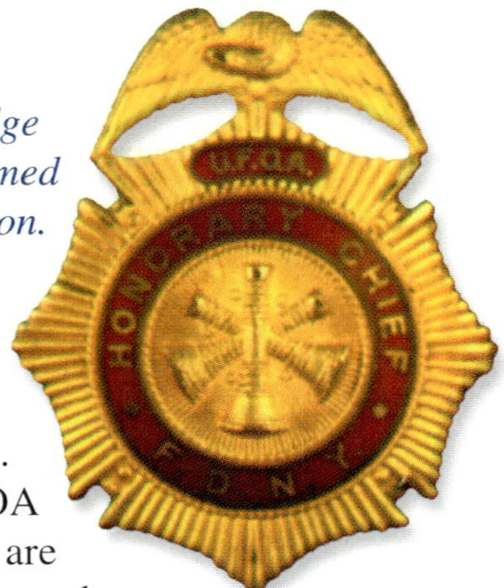

"Honorary Chief" badge presented by the Uniformed Fire Officer's Association.

- F.D.N.Y." in the outer ring. Badges made by the UFOA for presentation purposes are usually the same basic badge as the uniform badge with the New York City seal in the center and the words, "Uniformed Fire Officers Association" in red reverse enamel in the concentric ring.

Fire Marshals Benevolent Association

The Fire Marshals Benevolent Association (FMBA) was founded in 1964 for all Fire Marshals, regardless of rank. The goals of the FMBA are to set professional standards in the field of arson investigation and to work proactively with the Department's fire prevention initiatives. Their logo combines the two popular badges of the FDNY with a sun-ray style Fire Marshal's Maltese cross with eagle superimposed on a firefighter's-style Maltese cross. In the four arms of the larger cross are the letters "F" "M" "B" "A," respectively. In the concentric ring of the sun-ray cross in red reverse enamel are the words, "Fire Marshals Benevolent Association."

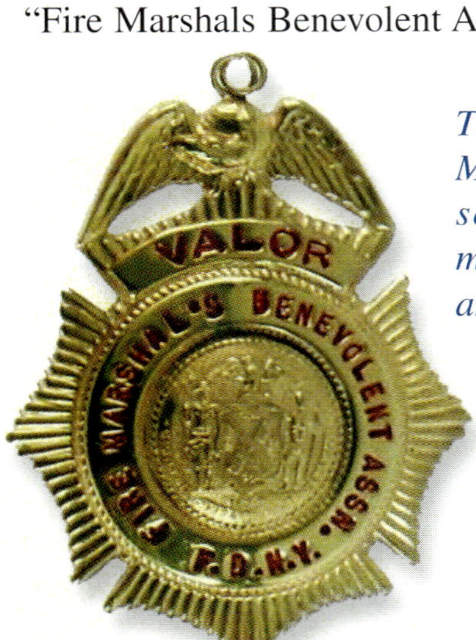

This badge from the Fire Marshal's ' Benevolent Association is actually the medal from the association's annual award for bravery.

Uniformed Pilots & Marine Engineers Association

The Uniformed Pilots & Marine Engineers Association (UPMEA) was founded in 1935 for uniformed members of the Department who have been promoted to the ranks of Pilot or Marine Engineer. All members are licensed Merchant Marine officers with a valid United States Coast Guard license. The objectives of this association are to improve the techniques and skills of its members in order to maintain the high quality marine fire protection in and around the New York City waterways.

The UPMEA utilizes the Marine Engineers and Pilots badge as its logo, superimposed on a large ship's anchor. In the central disk are the letters, "UPMEA" molded to fit and the motto, "In Unity There is Strength" in the concentric ring.

Fire Alarm Dispatchers Benevolent Association

The Fire Alarm Dispatchers Benevolent Association (FADBA) is a Local of the International Brotherhood of Electrical Workers, the labor organization representing Dispatchers and Supervising Dispatchers. It was founded in 1962. The logo is the standardized "Communications" badge that is issued to members in these positions. Die-struck badges with the FADBA name on it have been produced for officers of the association and have been affixed to presentation plaques for various awards.

Badge mounted on a plaque presented by the Fire Alarm Dispatchers Benevolent Association.

Fire Chiefs Association

The Fire Chiefs Association (FCA) is open to active, honorary, and retired Chief officers. The goals of the association are to protect the merit system and the general welfare of the Chief officers, and to unite all Chief offic-

NYC Fire Chiefs Association badge that was mounted on a presentation plaque.

Small lapel badge for the NYC Fire Chiefs Association.

ers through social, fraternal, and vocational cooperation. In December 1914, forty-eight Chief officers met to discuss establishing an association. Temporary officers were elected and plans were made to formally organize. The FCA was officially inaugurated in January 1915 as the "Chief Officers Endowment Association of the New York Fire Department." making it the oldest of the current line organizations. The name was changed to the Fire Chiefs Association in 1945. Its logo is a gold Chief's badge with the words, "Fire Chiefs Assn" in place of a rank and the date 1915 in the number plate.

Honorary Fire Officers Association

The Honorary Fire Officers Association (HFOA) is a fraternal organization that was founded in 1976. Its purpose is to support the FDNY with both personal service and financial support when requested. The financial support is utilized for projects that are not covered by municipal funding. As described in Chapter 12, appointment as an honorary officer in the FDNY is truly an honor and is only bestowed on a select group of individuals. Since 1994, membership in the HFOA is required for all honorary officers upon their appointment.

The logo of the HFOA is an interesting variation of the stanard FDNY officer's badge. It is a sun-ray Maltese cross with the official New York City seal in the center, not the version of the seal common to other badges. The seal is surrounded by two concentric circles. The inner circle bears the words, "Fire Department • City of New York." The outer circle bears the words, "The Honorary Fire Officers Association." The letters "FDNY" appear on the number tab.

Holy Name Societies

Two Holy Name Societies (HNS) are established within the Department. The HNS for Manhattan, Bronx, and Staten Island was formed in 1911. The HNS for Brooklyn and Queens was founded in 1922. Both groups are based on Catholic doctrine and provide spiritual camaraderie to their members who represent all active and retired ranks and positions of the Department. They are involved in numerous charitable and educational endeavors and they sponsor annual memorial and Communion masses for their members.

Various uses of the badge logo Holy Name Society, Holy Name Society of Queens and Brooklyn (above), and Holy Name Society of Manhattan, Bronx and Richmond (below.)

Both Holy Name Societies utilize the basic Maltese cross as their logo with the letter, "F.D.N.Y." in the upper arm and the names of the boroughs in the other arms. For the Manhattan and Bronx society, the letters "H.N.S." are in the lower arm. In the center of the badge is a silhouette of Christ with a halo and a crucifix.

Naer Tormid Society

The late Rabbi Edward Lissman, Chaplain Emeritus of the FDNY, founded the Naer Tormid Society (NTS) in May 1926 for Department members of the Jewish faith. Naer Tormid means *perpetual light* in Hebrew. Since its formation, the NTS has maintained headquarters at Riverside Synagogue. The Society symbol is a standard Maltese cross with the Star of David in the central disk and the letters "NTS" in its center.

The NTS bestowed honorary membership on Chief John J. McElligott in 1932. The badge is a pentagon, the standard FDNY officers' badge of the period. It is cast in solid gold and bears a presentation inscription on the back.

Honorary Member ribbon with badge from the Naer Tormid Society.

Lapel badge of the Naer Tormid Society.

Honorary Member badge presented to Commissioner McElligott by the Naer Tormid Society.

St. George Association

Organized on October 20, 1937, the St. George Association promotes spiritual development and social welfare amongst its members and fosters the religious spirit of the Protestant faith. Its logo is an FDNY firefighter's Maltese cross with the image of St. George slaying the dragon in the center surrounded by the words, "St. George Association - N.Y. Fire Department".

New York Fireman's Cycle Club

In 1896, Honorary Medical Officer Harry M. Archer established the New York Fireman's Cycle Club to promote fitness. Dr. Archer, who lived to eighty-five years of age, was a firm believer in the benefits of fitness. Aside from bicycle riding as a hobby or sport, it had practical applications for firemen of the day who used the two-wheeled vehicles to optimize their few short breaks from the continuous duty that was required at the time.

On December 11, 1896 the Cycle Club designated a uniform to be worn by members as well as a badge consisting of the standard fireman's Maltese cross with a bicycle wheel in the center. This continues to be the Club's insignia.

Original badge designed by Dr. Harry Archer for the NY Fireman's Cycle Club in 1896.

American Legion - Post 930

Many American Legion Posts across the country are affiliated with other fraternal or civic organizations. Post 930 is open to active and retired members of the uniformed force of the FDNY who are veterans of the United States armed forces. It was established shortly after the First World War in 1920.

Label badge from FDNY American Legion Post 930.

The insignia of the Post incorporates the two organizations that it represents with the national logo of the American Legion in the center of a Maltese cross. The left and right arms of the cross bear the words, "Post 930."

Columbia Association

The Columbia Association was founded in the early 1930's by Captain John Capillo of Engine Company 28. Its initial mission was to provide fire safety programs for residents of the lower East Side of Manhattan, many of whom were Italian immigrants who did not speak English. This area of the City was known for its overcrowded and unsafe tenement dwellings. Capillo, who spoke fluent Italian, carried the message of fire prevention to these inhabitants. The current association has broader objectives and is open to any members of Italian ancestry.

The logo of the organization incorporates the two predominant FDNY badge styles, namely the fireman's Maltese cross superimposed on the sun-ray Maltese cross with eagle. Within the central disk is a depiction of Columbus' ship, the Niña, encircled with the words, "Columbia Assoc."

A badge presented by the Columbia Association to Commissioner Patrick Walsh was made in the format of an FDNY officer's badge in gold. It bears his name in the upper panel and the words "Honorary Member - Columbia Association - FDNY" in the title ring. Also, the eagle

Custom-made badge presented to Commissioner Patrick Walsh as Honorary Member of the Columbia Society.

atop the badge was enhanced, giving it a more defined, three-dimensional appearance. The City seal, as on standard badges, is in the center in lieu of the association's traditional logo.

Emerald Society

Founded by Michael C. Donohue on March 17, 1956 (St. Patrick's Day), the Emerald Society is dedicated to the preservation of the literature, music, culture, language and traditions of Ireland. Perhaps one of the most visible activities of the Society is its bagpipe band, a central attraction in New York's annual St. Patrick's Day parade, the largest of these parades in the United States. The Emerald Society Bagpipe Band, formed in 1962, has received much acclaim.

The official emblem consists of a panoply of Maltese cross, shamrock, lyre, and sword bordered by upper and lower banners with the words, "Emerald Society."

The official badge authorized by the Department is manufactured for the Society by the C. G. Braxmar Company. This badge is only available to members of the Society. An unofficial badge, shown in Chapter 17, was speculatively made in 1997 by another manufacturer who offered it for general sale. It differs from the official badge as demonstrated by the shamrock applied to the center and the low-quality pin on the back.

Lapel badges issued by the Emerald Society.

Hispanic Society

The Hispanic Society, open to any uniformed or civilian employee of the Department who is of Hispanic origin or ancestry, was founded in 1962.

The symbol used by the Society is a Maltese cross with the letters "HS" in the upper and lower arms respectively. In the center disk is the image of a Spanish galleon. The cross sits on a double-layered banner with the words, "Hispanic Society F.D.N.Y.C."

Pulaski Association

The Pulaski Association, founded in 1960, is open to all members of the Department who are of Polish or Slavonic ancestry or whose spouse is of such heritage. It is affiliated with the Grand Council of Pulaski Associations. The association's official logo is a Maltese cross with a portrait of General Casimir Pulaski in the center.

Official FDNY sanctioned badge for the Emerald Society Pipe Band.

Membership card from the Pulaski Association.

The letters "FDNY" appear in the left and right arms of the cross. The words, "Pulaski Ass'n. Org 1960" are divided between the upper and lower arms.

A badge for the Pulaski Association was designed in 1998. It consists of a standard fireman's Maltese cross in silver. A Polish white eagle on a shield of red enamel is displayed in the center disk. This badge is only available to members of the Association.

Unofficial, prototype badge of the Pulaski Association.

Steuben Association

The Steuben Association is comprised of Department members who are of German birth or ancestry. Formed in 1963, its activities are patriotic, social, and charitable.

The symbol of the Steuben Association is a central FDNY Maltese cross displaying a portrait of Prussian Baron Freidrich Wilhelm von Steuben. The cross is surrounded by a laurel wreath encircled with the words, "Steuben Association Organized 1963."

Fire Square Club

The Fire Square Club is open to active and retired Department personnel who are members of good standing in any duly constituted Masonic Lodge. The Club was initiated by Battalion Chief Albert Reid. It was organized on January 26, 1921, making it one of the older line organizations. The purpose of the club is consistent with the purpose of the Masons which is based on fundamental commitments to charity and spirituality.

The Fire Square Club, formed in 1921, consists of FDNY members who are also members in good standing of the Masons.

The symbol of the Fire Square Club does not utilize the Maltese cross. It displays a fire helmet above a Mason's square, encircled by a hose with nozzle and coupling and the words, "The Fire Square Club of New York City, Org. 1921." A lapel pin with the Club's logo is shown here.

Vulcan Society

The Vulcan Society was founded in April 1940 by Battalion Chief Wesley Williams for African-American members of the Department. Its purpose is to meet the needs of its members in matters of social and economic welfare and general well-being. The Society is dedicated to the highest principles of patriotism and loyalty to the Department and the City.

The group's emblem is the FDNY Maltese cross with the letters "VS, Inc." in the center disk.

United Retired Firefighters Association

Perhaps the latest line organization is the United Retired Firefighters Association (URFA). For many years, retirees of the Department, particularly those living outside of the City, have formed local groups to maintain the camaraderie they knew when they were "on the job." But there was no formal structure tying the individual groups together other than a loosely structured "Grand Council of Retired Firefighters." In 1989, the Grand Council elected a President and in 1993 it became a corporation with the name, United Retired Firefighters Association. Currently, there are twenty-two chapters of the URFA nationwide. They include the Veteran Fireman's Association, the oldest organization for retired members, formed in 1884 (see Chapter 5). The URFA maintains offices at the same location as the UFA and publishes a newsletter, *The Retiree's Voice*. It offers benefits for retirees and serves as a strong, unified voice for its members. The organization represents both firefighters and officers who were represented by separate unions while on active duty. The URFA does not issue a badge to its members, however several of its officers have received courtesy badges, one of which is displayed here.

Logos of FDNY line organizations.

Badge of Jack Cunningham, Secretary-Treasurer of the United Retired Firefighter's Association (circa 1999.)

Lapel badge from the FDNY Boxing Club.

Logos of FDNY line organizations

CHAPTER 16

FDNY Related/Recognized Organizations

Although firefighting has been identified as the second most stressful job in America, (the presidency being number one), there has always been a certain romance associated with it. This romance is somewhat undefinable but is certainly the reason that many are drawn to the profession or the hobby. The New York City Fire Department's very long and colorful history has contributed to this sense of romance and attracted what is probably the largest following of devotees and collectors of memorabilia.

While the informal activity of "buffing" is officially discouraged by the Department, several buff organizations are officially recognized by the Department. As a symbol of membership and identity with the FDNY, most of these groups issue badges which are illustrated in this chapter.

Fire Bell Club of New York

On a tip from a police officer, twenty-two Buffs sought shelter from the cold in a loft near the quarters of Ladder 24 on February 3, 1939. Soon thereafter, the group rented the loft and established itself as the Fire Bell Club of New York (FBC). Since the Department banned such organizations in 1841, the Fire Bell Club was determined not to repeat mistakes of the past. Instead it worked to gain the respect and cooperation of the Department. The club's constitution describes its objectives as:

1. To perpetuate the memory and spirit of the men of the firefighting profession.

2. To preserve documents, photographs and relics pertaining to the history of firefighting.

3. To devise and give effect to measures fostering interest in and knowledge of the functions of the firefighting profession.

4. To maintain a clubhouse or clubrooms for the preservation of the various documents, photographs and relics pertaining to the firefighting profession and also for the use, enjoyment and entertainment of the members of the club.

Several of the original members of the Fire Bell Club.

Like most Buffs, FBC members frequently responded to the scene of greater alarms. Much to their credit, they did so in a manner that gained the respect and support of the Department. This was obvious in 1946 when Commissioner Frank Quayle authorized official Fire Department badges to members of the club. This has been the only such authorization given to non-departmental personnel, with the exception of Honorary Officers. The FBC badge is a gold sun-ray style Maltese cross without eagle. The letters, "F.B.C." are engraved in blue enamel on the top arm. In the center is the seal of the City of New York with a blue enamel background. In a concentric ring to the seal, are the words, "Fire Department City of New York" are displayed in blue enamel. There is a small disk on the lower arm where the member's number is engraved.

The only Department-sanctioned badge for an affiliated organization, the Fire Bell Club.

The Fire Bell Club remains active today, supporting many FDNY activities. Since 1984, it has awarded an annual medal for valor to a member of the Department. The club's headquarters office has moved twice since it original loft near Ladder 24. It is currently located at 150 East 23rd Street in Manhattan.

Third Alarm Association

The Third Alarm Association (TAA) was formed in 1945 by former members of the Fire Department Emergency Auxiliary Corps after they were mustered out of the civil defense initiative after the war. As the name suggests, the group responded to all third alarms. Their desire to be helpful and useful at the scene of fires was realized when the "Third Alarmers" organized a canteen to respond to the greater alarms, serving coffee and food to the firefighters. Since this task was logistically difficult on a city-wide basis, the TAA made arrangements with its colleagues in the 255 Club (see next section) to take over this responsibility in Brooklyn.

Early (left) and current (below) badges of the Third Alarm Association.

The TAA is still active today. Its members have expanded their involvement with the FDNY in many ways including support for the N.Y. Firefighters Burn Center and presentation of an annual medal to a member of the Department.

There have been two badge designs utilized by the TAA. The first was a silver Maltese cross with "Third Alarm" engraved in black enamel letters on the upper arm, the word "Association" similarly engraved on the lower arm, and an applied red enamel equilateral cross in the center. This badge had been issued since the group's initiation in 1945. The TAA's current badge is a sun-ray Maltese cross without eagle, in gold, bearing the letters "T.A.A." on the upper arm in the same manner as "F.B.C." on the Fire Bell Club badge.

Contemporary helmet of the Third Alarm Association.

255 Club

In 1941, a group of Buffs from Engine 255 (and Ladder 157) in Brooklyn were among the first to become auxiliarists and to take an active role with the companies. The Department directed that each company of auxiliarists should form associations. The 255 Club continued long after the end of the F.D. Emergency Auxiliary Corps, when they returned to traditional buffing. In 1946, the club broadened its participation in FDNY activities when it assumed responsibility from the Third Alarm Association for manning a canteen at greater alarms in their home borough of Brooklyn.

The 255 Club's badge is a standard silver-color Maltese cross with the words, "255 Fire" engraved with black enamel in a flat panel on the upper arm and "Club" in the lower. In the center is a freestanding, red-enamel, equilateral cross.

Signal 7-7 Club and Museum of Brooklyn

Since the days of the bell system for transmitting alarms to fire houses, the prefix signal "7-7" has signified a Borough Call (6th alarm) and a notification signal for Brooklyn and Queens.

The Signal 7-7 Club had, for many years, maintained their own museum at their headquarters on Bridge Street in Brooklyn. The club was disbanded shortly after the headquarters was severely damaged by a fire in 1980. Some of the collection was destroyed and the remainder was donated to the NYC and other fire museums.

The logo of the Signal 7-7 Club was a Maltese cross with the words, "7-7" and "Club" in the upper and lower arms respectively. In the center was the national symbol of Civilian Defense Auxiliary Firemen, which was intro-

Members of the 255 Club serving hot coffee to firefighters in 1947.

The 255 Club was originally the Auxiliary Corp association of Engine 255.

The old Signal 7-7 Club quarters in Brooklyn, in 1949.

duced in and used during WWII (see Chapter 11). Of the post-war clubs that grew out of the Auxiliary Corps, only the Signal 7-7 retained use of this emblem. No badges from the club were found and, based on information from former club members and badge manufacturers, it is believed that badges may have been made by individual members not the club as a whole. Based on the club's logo, a rendition of what such a badge would look like is shown here.

The Signal 8-8 Club owned a retired engine from the FDNY.

Though none were found to exist, this is the badge logo used by the Signal 7-7 Club and Museum of Brooklyn.

Signal 8-8 Fire Buff Club

The Signal 8-8 Club takes it name from the same source as the Signal 7-7 Club. "8-8" is the telegraphic alarm designation for Staten Island, the Borough of Richmond. The club, organized in 1959 and headquartered in Staten Island, was a long-time and proud owner of a C-85 Mack pumper, formerly FDNY Engine 158.

The Signal 8-8 Club is still active. Its badge consists of a sun-ray style Maltese cross without eagle. In a banner on the upper arm is the word "Member." The City seal is in the center and the words, "The Signal 8-8 Fire Buff Club, Inc." are engraved in the concentric ring. The member's number is engraved in a small panel in the lower arm.

Originally the Signal 6-6 Club, the Association of Auxiliary Firemen (later Firefighters) was another outgrowth of the WWII Auxiliary Corps.

Salvation Army Disaster Services

When the Uniformed Fireman's Association purchased a canteen to provide refreshments for members at multiple alarm fires, it was staffed by volunteers from the Salvation Army. In 1958, the Department discontinued operating the canteen. Several Buff organizations, (e.g., Third Alarm Association, 255 Club,) had already begun to provide canteen services. At that time, the Salvation Army purchased its own canteen vehicles and they continue to operate two of them.

A badge for the Salvation Army Disaster Services is shown. It is a gold-tone badge with red enamel lettering. It displays the initials, "S.A.D.S.N.Y." and the words, "Fire Department, City of New York". However, this badge was not issued by either the Department or the Salvation Army. It was most likely made by and for a volunteer.

Auxiliary/Volunteer Firefighters Association

Details regarding the Auxiliary/Volunteer Firefighters Association are sketchy. It was apparently formed sometime after the Second World War by members of the Emergency Auxiliary Corps and was, at times, also known as the Signal 6-6 Club. Signal 6-6 is the telegraphic signal for Manhattan, the group's area of activity. Around 1947, the Association published a list of all alarm box numbers and locations throughout the City. The Association is no longer in existence.

The badge that was used by the Association consisted of a sun-ray Maltese cross. Officers had eagle-topped badges with their rank in a banner held in the eagle's beak.

The Salvation Army Disaster Services canteen has been a welcome sight to NYC firefighters for many years

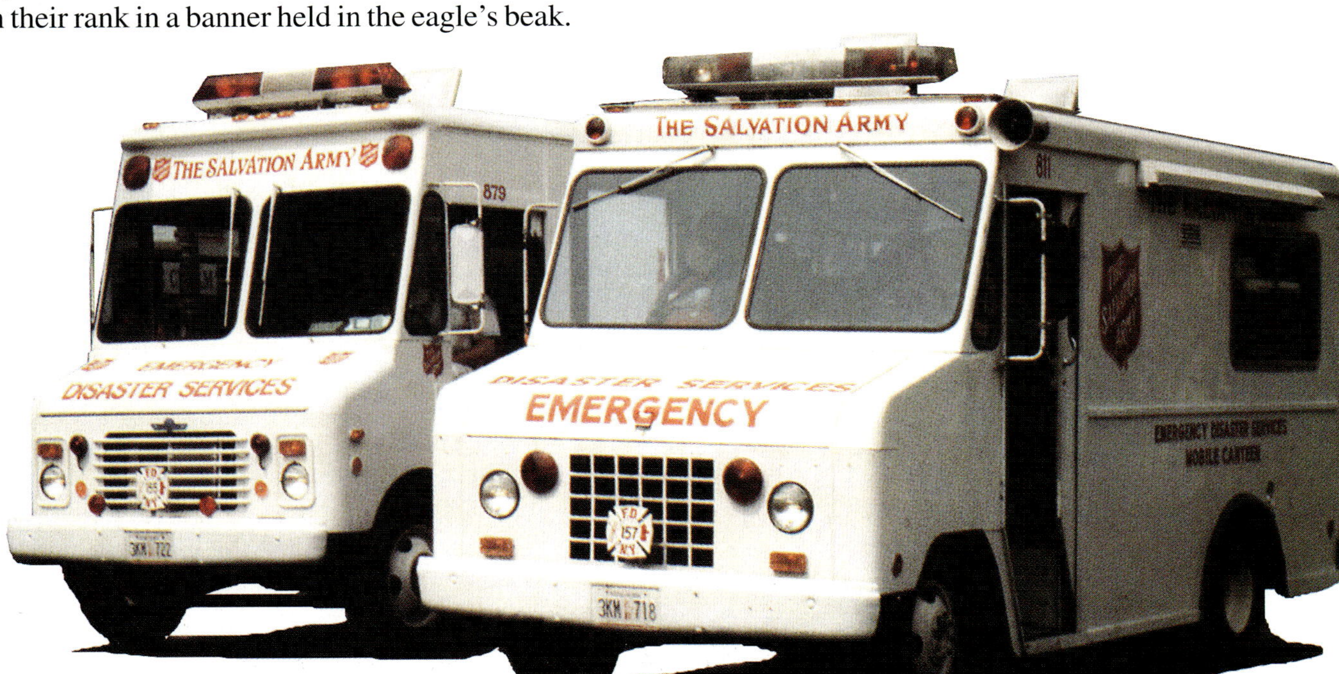

Canteen units operated by the Salvation Army Disaster Services. (Photo courtesy of E. M. Koehler, Jr.)

CHAPTER 17

Unique And Unusual Badges

Horses

Horses were introduced into the Department in 1832 and made their last run with Engine 205 on December 20, 1922. The presence of horses in the Department required several supporting positions, such as blacksmiths and hostlers, some of whose badges are shown in Chapter 10.

As loved though they were, fire horses were not recognized by the Department with an official badge. They were, however issued a Departmental number. This number was stamped into a brass tag that was suspended from the neck. These were discontinued on March 24, 1910 and the number was branded into the horse's hoof. None of these number tags are known to exist. Photographs of some fire horses show a number suspended from a face piece, or drop. The number is however, that of the company not the horse. Finally, on the harness of some horses was a rosette bearing the New York City seal. Though not a badge, these are nonetheless considered by collectors to be meaningful relics from the horse-drawn era.

Ironically, Police Department horses are issued a badge which it is affixed to their saddle blanket. Unfortunately, it is too late to undo this injustice to the firefighters' equine colleagues who performed their job so courageously and vigorously.

Though not truly a badge, this is one of the last remaining artifacts worn by NYFD horses.

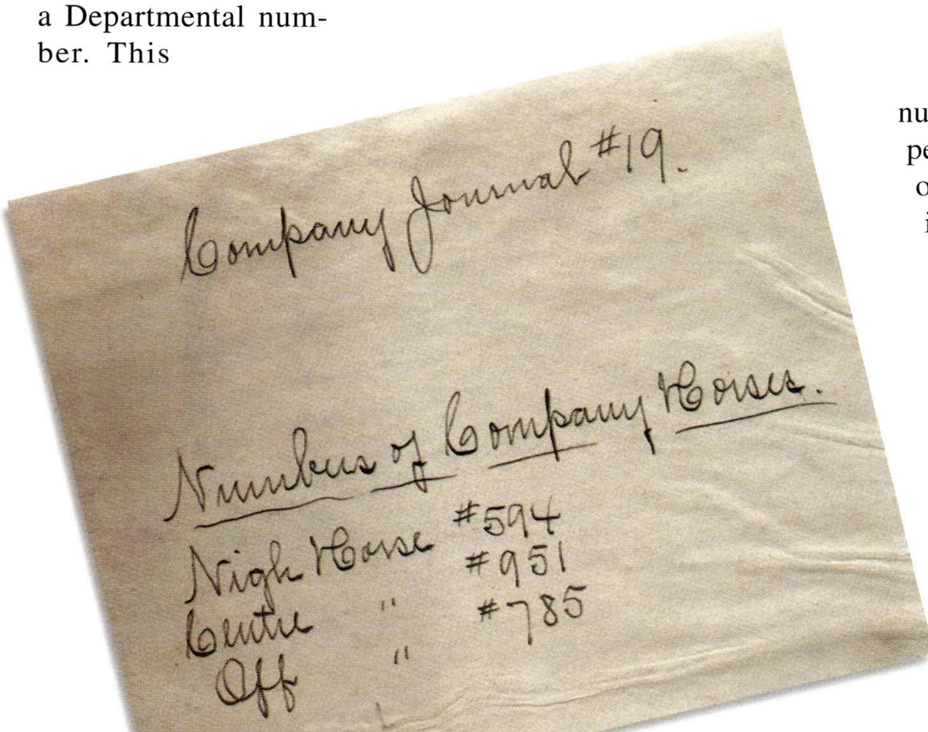

Company Journal #19.

Numbers of Company Horses.

Nigh Horse #594
Centre " #951
Off " #785

Entry in 1912 journal of Ladder Company 34 of the company's horses. The term "nigh" means the horse on the left and "off" means the horse on the right, in a three-horse rig.

Crystal Palace Fire

Horses of Engine Company 169 wearing face drops displaying the company number.

Among the "firsts" for New York City was the hosting of the first World Exposition, or World's Fair. To accommodate this 1853 event, a grand structure of glass and iron was constructed in midtown Manhattan in the likeness of its predecessor and namesake in England, the Crystal Palace. The exposition was held from July 14, 1853 to November 1, 1854, after which the building remained and was used as a convention center. On October 5, 1858 a fire consumed the facility in a reported time of only 15 minutes. Five pieces of apparatus from the New York Fire Department were lost in the fire. They were the hose wagons of Eagle Hose 1 and Croton Hose 6, the ladder truck of Mutual Hook & Ladder 1, and the engines of Gotham Engine 16

Commemorative badge of the Crystal Palace fire in 1858 that destroyed five pieces of NYFD apparatus.

The Crystal Palace being consumed by fire on October 5, 1858.

and Pacific Engine 28. A detailed account of this fire can be read in, *Our Firemen* by Augustine Costello.

To commemorate this historic fire and the loss of the equipment, a souvenir badge was struck. It is not known who sponsored this badge or when. The badge is in the shape of a helmet front, with the words, "Pacific 28 - Burnt in the Crystal Palace Fire October 5th 1858."

1887 New York to San Francisco

During the 1800's and into the early 1900's, it was common for fire departments to host conventions or other gatherings for visiting firemen. New York firemen had envisioned a transcontinental journey that would take them to fire departments across the country. This dream became reality when, on September 5, 1887, one hundred eight members and officers of the Veteran Firemen's Association of New York City embarked on their journey amid a great deal of fanfare. Traveling by rail to their ultimate destination of San Francisco, they visited fire departments in Omaha, Salt Lake City, Sacramento, Denver, Kansas City, St. Louis, Louisville, Cincinnati, and Niagra Falls. The members returned to New York on October 1, 1887.

Souvenir badge of the coast-to-coast trip of NYC firemen in 1887.

Many souvenirs were created to commemorate this historic trip. They included a large photograph of the veterans parading through the streets of New York on their way to the train that carried them across the country, a journal printed after the event, and a small badge. The badge is frontpiece-shaped and is made of gold with blue enamel. It has ornate flourishes and a centerpiece showing a hook, ladder, and helmet as well as two shaking hands.

Eligible Firemen

In 1930, firemen were working 84 hours per week in a two-platoon system. The movement to institute an 8-hour shift, three-platoon system generated heated debate between the affected factions. Those in favor of a three-platoon system included men who had taken the civil service examination to become firemen and were waiting to be called. Clearly, the addition of another platoon would result in the hiring of men from the list. Men on the list were organized by the Uniformed Fireman's Association

New York firemen embarking on their sojourn cross country to San Francisco.

Political cartoon appearing in 1937 after passage of the Three-Platoon Bill.

and sent out into their communities to muster support for passage of the three-platoon system.

The union made badges for the men to wear while they went door-to-door to solicit support. The badge was a small silver Maltese cross based on the same die as UFA courtesy badges (see Chapter 15) but the center disk has the word "Eligible" and the dates 1931-1935. The number manufactured is unknown and few exist today.

A very rare badge issued to Eligible Firemen during the bitter struggle to establish a third platoon.

125th Anniversary

Although designed and sold by a private seller, a commemorative badge for the FDNY's 125th Anniversary was issued in 1990. It is a pinched shield shape and shows a steamer on cobblestones in beautiful detail. The badge is made with the Blackinton "Rose-Gold" finish giving it a bronze appearance. It has blue reverse enamel with the wording, "Metropolitan Fire Dept. - 125th Anniversary - FDNY." The badge was produced as a limited edition of 360. It was delivered in an attractive blue suede box with white satin lining imprinted in gold with the words, "Fire Department of the City of New York - 125th Anniversary." Each badge has a serial number on the reverse.

Badge struck to commemorate the 125th Anniversary of the paid FDNY. Designed by Steve Scher.

St. Patrick's Day

In 1997, United Insignia, one of the suppliers of badges to the FDNY, created and sold a unique badge to celebrate St. Patrick's Day. It is worn by members of the Department who march in the country's largest St. Patrick's Day parade down Fifth Avenue past St. Patrick's Cathedral. The badge is a standard FDNY Maltese cross with a large green enamel shamrock in the center. In 1999, an officer's version sunray Maltese cross with eagle was produced in various ranks.

1997 creation of United Insignia for NYC firefighters celebrating St. Patrick's Day.

Although these badges in no way conform to the Department's Uniform Equipment Specification, the Department seems to tolerate the violation on this festive occasion. Given the party atmosphere of the day and the post-parade celebrations, the number of lost official badges is minimized by using this substitute.

Vehicle Badges

In 1865, when volunteer companies were taken over by the paid Department, company names were eliminated. Company numbers continued to be used and, in most cases, were reassigned. The number was usually prominently displayed on the apparatus. As specialty companies and apparatus were introduced, they too received designated numbers. In 1923, the Department Shops began to affix a Maltese cross badge on certain vehicles. Engine, ladder, rescue and other companies directly involved in firefighting did not receive the new badges. They were affixed to the cars of Chiefs, high-ranking officials and specialty vehicles such as ambulances, fuel trucks, mask service units, oxygen therapy units, etc.

The vehicle badges were made of cast aluminum measuring 8 1/2 inches square. They were painted white with red on raised areas. The badge/vehicle number was made of whatever material was utilized at the time, (e.g., metal, plastic and, in later years, Scotchlite®.) A variation of the standard vehicle badge has gold stars affixed in the top arm for high-ranking Chief officers. This design was introduced in 1971, when stars were added to certain breast badges.

In the early 1970's, most metal vehicle badges were replaced with a Scotchlite® decal. Metal badges continued to be affixed to the cars of the Commissioner, Chief of Department and other high-ranking officials with vehicle identification numbers under 100. This practice continues today.

Vehicles designated as fire apparatus are not required to display a standard New York State license plate. In the 1970's, a Maltese cross license plate was introduced. It is

Early vehicle badge with metallic, riveted number

Contemporary vehicle badge with reflective vinyl number.

Badge mounted to vehicle of the Chief of the NY Fire Patrol.

Vehicle badge style introduced in 1971 with stars to indicate assignee's rank. This example is for the Bronx Borough Commander.

Scotchlite® reflective vehicle badge decal.

Vehicle badge #1 issued to the Fire Commissioner.

License plate-style vehicle number plate with Maltese cross.

of a size equivalent to a standard vehicle license plate, has a red Scotchlite® background with the outline of an FDNY Maltese cross to the left and the vehicle number in white Scotchlite® to the right. A variation of this displays gold stars in lieu of the vehicle number for high-ranking Chief officers.

Traditionally, the Commissioner has always been assigned Car 1 and the Chief of Department is assigned Car 5. As staff sizes have changed, vehicle numbers have changed as well. (The Chief of Department is now assigned vehicle badge number 3.) With the advent of vehicle radio communications, the vehicle number on the badge is also used as the radio identifier for the car and its occupant.

A rather unusual type of vehicle badge is in the collection at the New York City Fire Museum. It is a Maltese cross of eight inches diameter and three inches thick. On the front of the badge are holes that allow light from a bulb in the box to be emitted. Wording on the badge is different from the standard vehicle badge. It has the word,

Unusual vehicle badge with back-lighting of unknown origin or use. Currently in the collection of the NYC Fire Museum.

"FIRE" in the center and the letters "NYC" in the lower arm. The badge number is in the upper arm. Badge numbers 3 and 5 are in the collection which, given their age, would have been assigned to a Deputy Commissioner and the Chief of Department. Two leading authorities on FDNY vehicles believe that these badges were never issued or used by the Department. Their history is unknown.

The first Department ambulance was assigned a badge simply with the letter "A" in the center. Later ambulances had the letters, "AMB" in the center along with the ambulance number. Canteen kitchen vehicles operated by the Department (see Chapter 16) had a unique vehicle and telegraph designator, specifically 9-94, from 1946 through 1958. The telegraph signal "9" represented a request for special apparatus. The number "94" was chosen to represent the Uniformed Firefighters Association (UFA). The UFA is Local 94 of the International Association of Firefighters and it was the UFA that donated the canteen kitchen vehicle in 1946.

Vehicle badge authorized for the Department canteen.

Vehicle badge issued to a Department Medical Officer.

Some large metal badges that appear to be vehicle badges are not. These are roughly the same size as the standard vehicle badges but are cast in bronze instead of aluminum and are not painted. On many, the central disk is open. These badges were typically presentation pieces fitted with either a clock or a photo in the center disk. Another badge that is easily mistaken for an official FDNY vehicle badge is one with a number in the center, a name in the top arm and "FDNY" in the bottom arm. A very limited number of these badges had been made as house numbers. The number is the address of the house and the name is that of the resident, a member of the Department.

FDNY Maltese cross picture frames.

FDNY vehicle badge personalized for use as a house number plaque.

Vehicle badge issued to Second Deputy Commissioner Harry M. Archer, MD.

Some Honorary Chiefs are authorized to display an FDNY Maltese cross vehicle badge. However, the badge is modified so that the letter "H" appears in the upper arm and the letters "FDNY" in the lower. Numbers are assigned in a manner similar to standard vehicle numbering.

Of note is vehicle badge number 3, shown here. This badge was affixed to the car assigned to Dr. Harry Archer during his tenure as Second Deputy Commissioner.

The badge from Engine 3 of the 1964 World's Fair Fire Department is shown, although it is not from an FDNY vehicle. Unlike the 1939 New York World's Fair

Vehicle badge issued to some honorary officers.

Vehicle badge from the 1964 New York World's Fair Fire Department.

The Fireboat Harvey with a Maltese cross visible on its stack.

The addition of badges on fireboat stacks appeared in an article in WNYF magazine in July 1956.

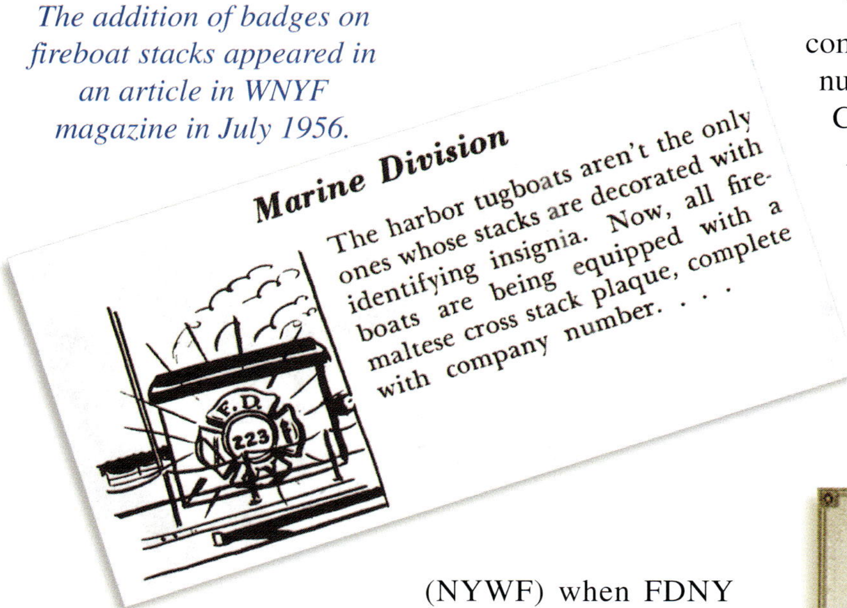

Marine Division

The harbor tugboats aren't the only ones whose stacks are decorated with identifying insignia. Now, all fireboats are being equipped with a maltese cross stack plaque, complete with company number. . . .

(NYWF) when FDNY vehicles and personnel were stationed at the fairground, Pinkerton's International Detective Agency provided fire services at the 1964 NYWF. The badges from these fire brigades are presented in Chapter 21.

The Department's water-borne vehicles, its fireboats, were also given badges in 1956. An FDNY Maltese cross bearing the company number was painted in white and red on the vessel's smoke stack. On smaller craft, a reflective decal badge has been used on the sides of the hull.

Private Companies

Several New York City businesses that manufacture fire department supplies have produced badges that are sought after by collectors.

The late C. G. Braxmar.

Souvenir badge, distributed by Braxmar, one of the oldest and well-known producers of FDNY badges.

The first is from the C. G. Braxmar Company. This company has produced many badges for the FDNY and numerous other departments in and out of New York City. Many of the badges in this book are from Braxmar. A small badge made of nickel was distributed as an advertising promotion. The company was founded in 1879 by Charles Braxmar. Unfortunately, one evening in 1914 he investigated the odor of gas in his home with a candle. The explosion was fatal. The company eventually came under the ownership of one of his employees, Mr. John O. Veit, whose granddaughter owns it today.

EDWARD F. CROKER
Ex-Chief N. Y. Fire Dept., President and Treasurer

TELEPHONE 7911-7912 BRYANT

The Croker National Fire Prevention Engineering Company

EXECUTIVE OFFICE

N. W. Cor. Fifth Avenue and 46th Street, New York

Our employees are all retired officers and members of the New York Fire Department, neatly uniformed, under our personal supervision and thoroughly experienced in all matters from a fire prevention and extinguishing standpoint.

Our inspection and drill service provides for:

A daily, weekly, semi-monthly or monthly inspection of all premises wherein our inspectors examine your premises from cellar to roof, and furnish you with written reports regarding the existence of any hazard, with proper recommendation for the removal of same.

We also organize your employees into private fire brigades and instruct them as to the use and handling of fire appliances, also what to do in case of fire and HOW TO ACT TO PREVENT FIRE.

We drill your employees, without inconvenience or loss of time to assure them retaining their presence of mind and acting with coolness and judgment in case of fire, accident or other danger.

Section 775 A, Chapter 458 Laws 1912, require FIRE DRILLS in FACTORIES, INSTITUTIONS, etc. Are you complying with the Law?

Uniformed retired firemen furnished by day, week or month for all purposes.

Fire Prevention orders efficiently and economically handled for Owners, Agents and Tenants. Automatic Sprinklers, Standpipes, Fire Alarm System, Fire Hose, Extinguishers and approved appliances of all kinds designed and installed. Plans and specifications for new or altered buildings, prepared or examined for owners, lessees, architects, to comply with various Department requirements as well as reducing fire hazard and obtaining lowest insurance rates.

All matters arranged and conducted under the personal supervision of ex-Fire Chief Edward F. Croker.

1912 advertisement for retired-Chief Croker's fire prevention company.

The Croker Fire Prevention Corporation, as its name implies, was owned by the famous FDNY Chief of Department, Edward Croker. It was established in 1911 when the Chief retired from the Department. The company sold fire protection supplies, most notably soda/acid fire extinguishers and fire prevention services. The Chief ran the company like his own fire department. All the employees were former FDNY members who wore uniforms, including a hat with a badge shown here. The company also had a promotional give-away but, unlike Braxmar, it was not a badge. It was a book of matches. Perhaps the Chief had a sense of humor.

Retired Chief of Department Edward Croker ran his company like a private fire department including requiring the wearing of uniforms and this badge.

Badges used by private alarm companies for admittance through fire lines.

Sometime after the introduction of street fire alarm boxes in New York in 1851, private alarm companies installed fire alarm devices in buildings throughout the City which were connected to the FDNY alarm system. The alarm companies maintained the private systems and ensured that they interfaced properly with the Department's systems. Upon an alarm, the company would dispatch someone to the scene. For identification, these people were issued badges by their company, which were recognized by police and fire officials. There appears to have been no standard Departmental specifications for these badges. Several badge styles have been found and some examples are shown here. One notable alarm company badge displays the words, "Admit to Buildings." This is contrary to all other fire lines badges which prohibited entrance (see Chapter 13.) However, the workers of the alarm companies needed access to the building in order to reset or shut down the activated alarm system.

Souvenir badge given out by the Fabric Hose Company, located at 5 Barclay Street, NYC circa 1879 -1890.

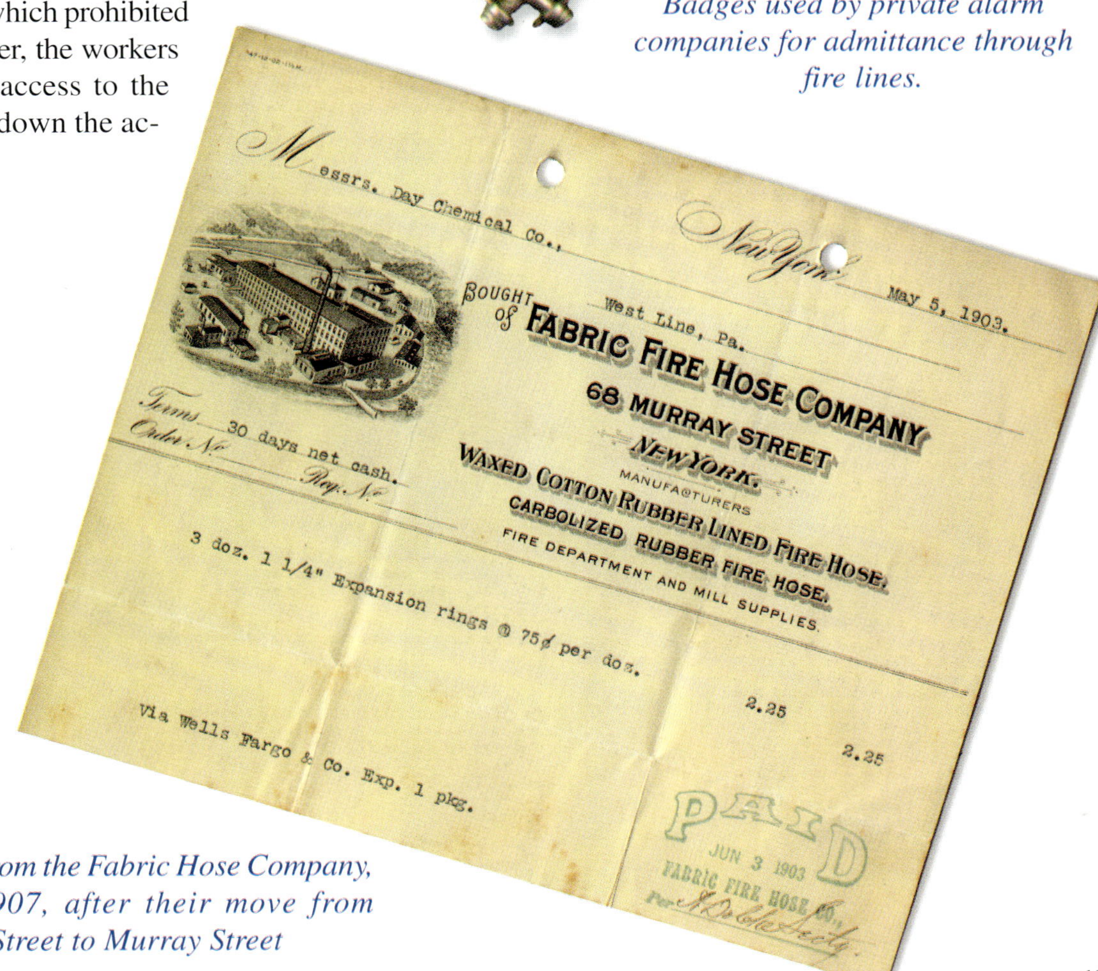

Messrs. Day Chemical Co.,

West Line, Pa.

New York May 5, 1903.

Bought of FABRIC FIRE HOSE COMPANY
68 MURRAY STREET
NEW YORK.
MANUFACTURERS
WAXED COTTON RUBBER LINED FIRE HOSE.
CARBOLIZED RUBBER FIRE HOSE.
FIRE DEPARTMENT AND MILL SUPPLIES.

Terms 30 days net cash.
Order No. Reg. No.

3 doz. 1 1/4" Expansion rings @ 75¢ per doz. 2.25

Via Wells Fargo & Co. Exp. 1 pkg. 2.25

PAID
JUN 3 1903
FABRIC FIRE HOSE CO.

Invoice from the Fabric Hose Company, dated 1907, after their move from Barclay Street to Murray Street

Although now known for being the financial hub of the world, lower Manhattan around Wall Street was home to many manufacturing companies. The Fabric Fire Hose Company was located at 5 Barclay Street circa 1879 - 1890, and later at 68 Murray Street and 127 Duane Street. One of their promotional give-a-ways was in the form of a badge. Perhaps the most interesting feature of the badge is its belted tiger logo of Americus Engine 6. It is not known if there was any connection between The Fabric Hose Company and Engine 6.

Juvenile

The FDNY, like most fire departments, has fire prevention programs for youngsters. Through the years, a number of educational items have been developed to promote fire prevention, such as comic and coloring books. Although the FDNY has made use of a variety of fire prevention materials targeted at children, including distribution of the National Fire Protection Association's, "Sparky's Junior Fireman" badge and materials in the 1960's, it does not appear that a special badge from the FDNY was ever manufactured for children.

In 1934, a new fire service publication was introduced for the New York Fire Department called *Fire Lines*. It offered a Junior Fire Lines Club for youngsters. Members received a small badge in the shape of a Maltese cross with the words, "Fire" and "Lines" in the upper and lower arms respectively and "Junior" in the center. The publication was short-lived, as was the club.

In the 1960's, the Department initiated a community relations program entitled, "Firefighters Devoted to Neighborhood Youth", the acronym is FDNY. A celluloid badge bearing an FDNY Maltese cross with the City seal in the center, in white on a red background, was distributed to children as part of this program.

The New York City Fire Museum attracts approximately 10,000 visitors per year of all ages. While some adults might like to take home some of the real items on display from the collection, youngsters have the pleasure of taking home a souvenir badge. Two such badges from the museum's gift shop are shown here. They are in the shapes of the Department's two current primary badges, namely the fireman's Maltese cross and the officer's sun-ray Maltese cross with eagle. They are stock badges from a premium company with the title "Junior Firefighter." The words, "NYC Fire Museum" are imprinted where a fire department name would typically be placed.

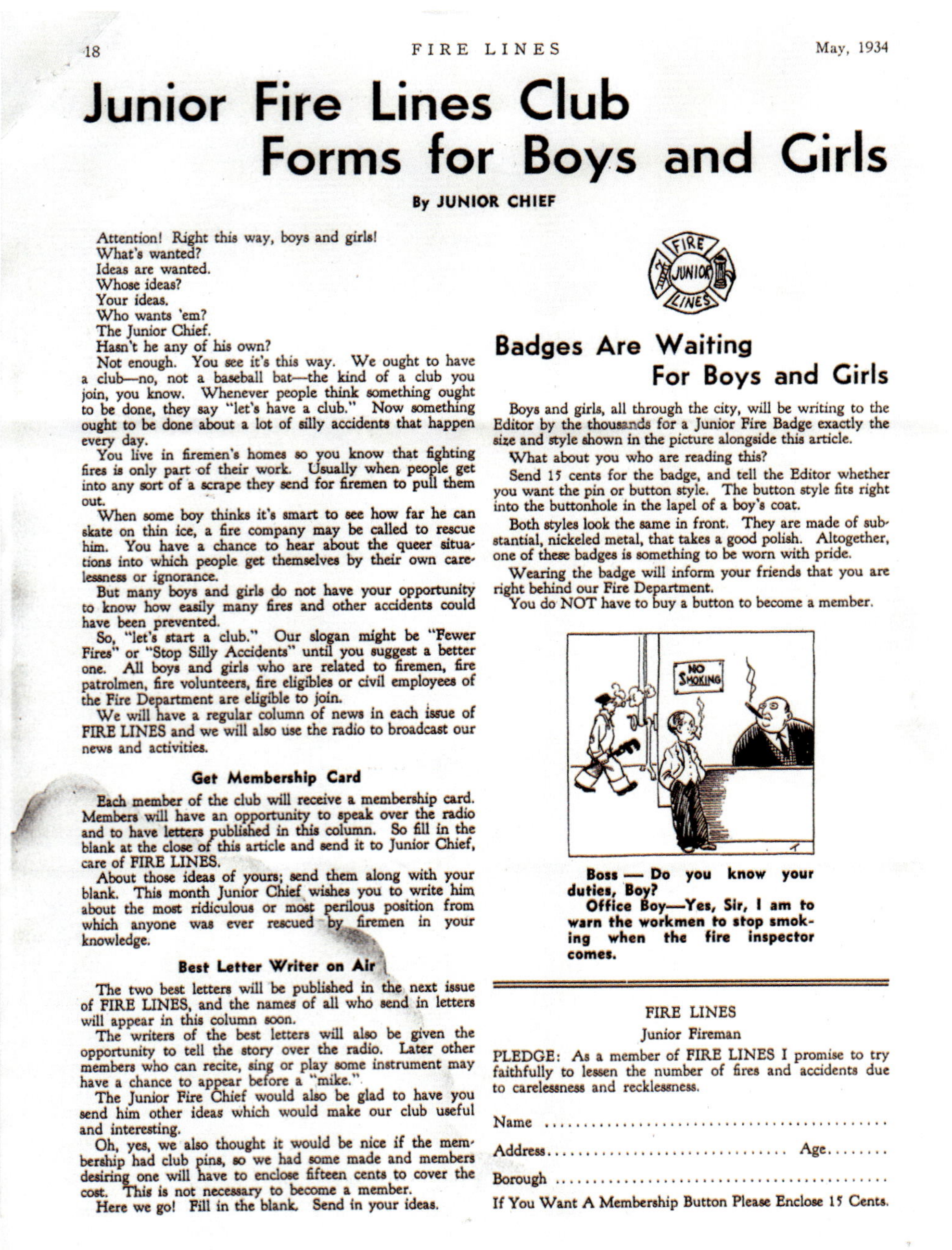

18 FIRE LINES May, 1934

Junior Fire Lines Club Forms for Boys and Girls

By JUNIOR CHIEF

Attention! Right this way, boys and girls!
What's wanted?
Ideas are wanted.
Whose ideas?
Your ideas.
Who wants 'em?
The Junior Chief.
Hasn't he any of his own?

Not enough. You see it's this way. We ought to have a club—no, not a baseball bat—the kind of a club you join, you know. Whenever people think something ought to be done, they say "let's have a club." Now something ought to be done about a lot of silly accidents that happen every day.

You live in firemen's homes so you know that fighting fires is only part of their work. Usually when people get into any sort of a scrape they send for firemen to pull them out.

When some boy thinks it's smart to see how far he can skate on thin ice, a fire company may be called to rescue him. You have a chance to hear about the queer situations into which people get themselves by their own carelessness or ignorance.

But many boys and girls do not have your opportunity to know how easily many fires and other accidents could have been prevented.

So, "let's start a club." Our slogan might be "Fewer Fires" or "Stop Silly Accidents" until you suggest a better one. All boys and girls who are related to firemen, fire patrolmen, fire volunteers, fire eligibles or civil employees of the Fire Department are eligible to join.

We will have a regular column of news in each issue of FIRE LINES and we will also use the radio to broadcast our news and activities.

Get Membership Card

Each member of the club will receive a membership card. Members will have an opportunity to speak over the radio and to have letters published in this column. So fill in the blank at the close of this article and send it to Junior Chief, care of FIRE LINES.

About those ideas of yours; send them along with your blank. This month Junior Chief wishes you to write him about the most ridiculous or most perilous position from which anyone was ever rescued by firemen in your knowledge.

Best Letter Writer on Air

The two best letters will be published in the next issue of FIRE LINES, and the names of all who send in letters will appear in this column soon.

The writers of the best letters will also be given the opportunity to tell the story over the radio. Later other members who can recite, sing or play some instrument may have a chance to appear before a "mike."

The Junior Fire Chief would also be glad to have you send him other ideas which would make our club useful and interesting.

Oh, yes, we also thought it would be nice if the membership had club pins, so we had some made and members desiring one will have to enclose fifteen cents to cover the cost. This is not necessary to become a member.

Here we go! Fill in the blank. Send in your ideas.

FIRE
JUNIOR
LINES

Badges Are Waiting For Boys and Girls

Boys and girls, all through the city, will be writing to the Editor by the thousands for a Junior Fire Badge exactly the size and style shown in the picture alongside this article.

What about you who are reading this?

Send 15 cents for the badge, and tell the Editor whether you want the pin or button style. The button style fits right into the buttonhole in the lapel of a boy's coat.

Both styles look the same in front. They are made of substantial, nickeled metal, that takes a good polish. Altogether, one of these badges is something to be worn with pride.

Wearing the badge will inform your friends that you are right behind our Fire Department.

You do NOT have to buy a button to become a member.

NO SMOKING

Boss — Do you know your duties, Boy?
Office Boy—Yes, Sir, I am to warn the workmen to stop smoking when the fire inspector comes.

FIRE LINES
Junior Fireman

PLEDGE: As a member of FIRE LINES I promise to try faithfully to lessen the number of fires and accidents due to carelessness and recklessness.

Name ....

Address .... Age ....

Borough ....

If You Want A Membership Button Please Enclose 15 Cents.

Recruiting advertisement in Fire Lines *magazine for the Junior Fire Lines Club.*

Replica of the badge issued by the "Junior Fire Lines Club" sponsored by the UFA in 1934.

Celluloid badge from the FDNY Community Youth program in the late 1960's.

Juvenile badges from the gift shop of the NYC Fire Museum in the shapes of the two standard FDNY badges.

The New York City Fire Museum is housed at 278 Spring Street in a building that was formerly the quarters of Rescue 1, Engine 30, Satellite 2 and the Medical Division.

CHAPTER 18

"Saver", Courtesy, Fraudulent, Unauthorized, Reproduction And Prototype Badges

When a firefighter is issued a badge, it is his or hers until retirement or promotion. Loss of a badge can result in a fine or temporary suspension imposed by the Department. Manufacture of unauthorized badges is a violation of the City Charter. But, when there is a will, there is a way. In the past, underground sources have supplied firefighters with a spare badge which the firefighter would use on a day-to-day basis, keeping the original secure. More recently, spare or "saver" badges have become easily obtainable, though proper Departmental identification must be presented in order to have one made. They are currently being struck from the same die as the authorized badges and are, for all intents and purposes, indistinguishable from the officially issued badges.

While it is a Departmental violation for a firefighter to have an unauthorized badge, possession of one by anyone else can lead to charges of criminal impersonation. As stated in Chief Engineer Howard's report of 1858, possession of a FDNY badge can give access within fire lines to *"thieves and other dangerous persons ...where... they can perform their nefarious calling."* In 1965, an individual posing as a fireman and possessing a badge entered firehouses and burglarized them. In order to add an additional level of security and to be more consistent with modern-day identification tools, the Department now issues photo identification cards to all uniformed and civilian personnel.

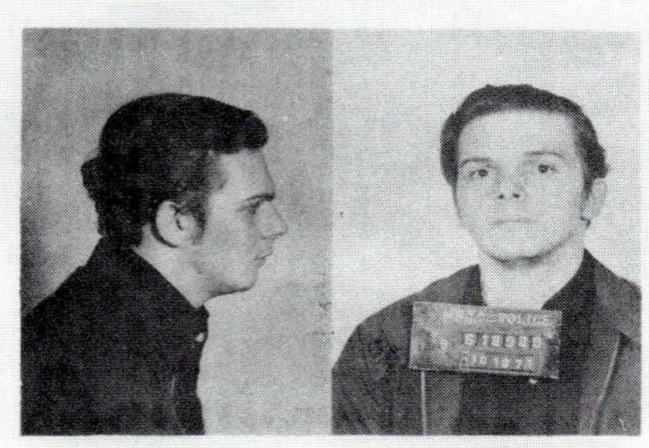

EDWARD BRUNNER
Date of Birth 7/14/49
5'8" – 160 lbs.
Dark brown wavy hair, light blue eyes, fair complexion, slim build, smooth talker.

The above individual has been arrested 10 times in the past 7 years – six times for impersonating a fireman. He had a fireman's badge in his possession at the times of five arrests. He has been charged with Trespass in firehouses on three occasions (Eng. 204, Eng. 310 and Eng. 323). He has also been charged with impersonating a Police Officer, theft of Police Walkie-talkies, theft of a gun on two occasions, burglary and other crimes.

He has been found to have a firehouse key in his possession and is suspected of thefts from other firehouses. He has been known to wear a fireman's uniform. He may say that "he is on the job", and use an alias.

He has recently been accompanied by another male, and on one occasion an Oriental. HE MAY BE ARMED.

EDWARD BRUNNER is NOT to be permitted on any Fire Department property should he present himself. If Brunner should be found in quarters, contact Chief Fire Marshal Edwin H. Sheppard at once at 566-2152 during business hours, and at Rh4-1000 at all other times.

By order of, ROBERT O. LOWERY,
FIRE COMMISSIONER

KEEP POSTED UNTIL FURTHER NOTICE

Wanted poster for a man using an FDNY badge to gain entry to fire houses around the City to commit burglary.

Unauthorized copies of FDNY badges made to resemble their legitimate counterparts.

Many unauthorized badges are commissioned by FDNY members themselves. These badges are of great interest to collectors of fire memorabilia and Buffs. There is a market for these items and FDNY helmets and badges draw a premium. Like any other supply-and-demand situation, the market for FDNY badges invites low-cost copies to be sold as originals. Depending on the manner in which they are made, bogus badges may or may not be easy to identify.

Many copies are made by a casting rather than a stamping process. As such, the features of the badge are typically much less distinct than on the original. Also, since a force is not used from behind the badge, the reverse will typically be very vague and uneven. This is probably the simplest, most obvious difference.

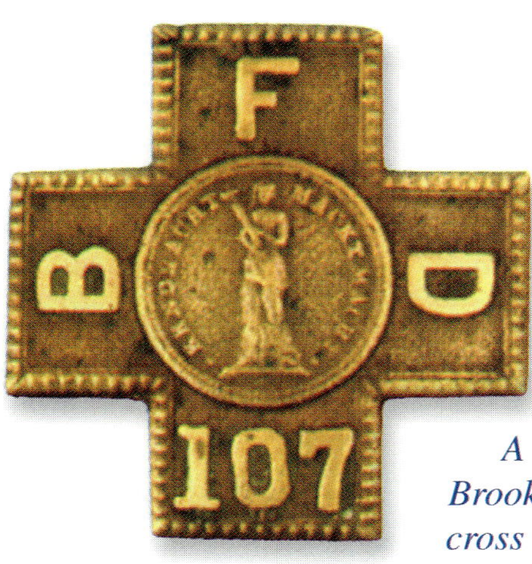

A reproduction of a Brooklyn Fire Department cross made from a casting.

Evidence of bogus badges including a single star applied to a Commissioner's badge and tell-tale backs of badges from castings.

Several decades ago, when no manufacturer would produce a badge that was an exact copy of an FDNY Maltese cross, copies were obvious. However, these copies became useful when "saver" badges were introduced to protect original badges from theft while companies were out of quarters. Three saver badges are illustrated here. As can be seen, several variations were made during this period. One clearly possesses all the characteristics of an

Saver badges made by members to use in lieu of their official badge.

Saver badge for a member of the New York Fire Patrol.

official badge, such as the "F.D." "N.Y." and issued number. The other example, has "H.L." in the upper arm, "F.D.N.Y." in the lower, and the company number in the center. This badge belonged to a member of Hook & Ladder Company 25.

A badge with the letters, "F.P.N.Y." in the shape of a Maltese cross is shown. The New York Fire Patrol does not employ the standard Maltese cross as a badge (see Chapter 6). Given the similarity of this badge to the two other saver badges described and shown, it is believed that this too is a saver, made by a member of the New York Fire Patrol.

Many older, unofficial badges varied from the official ones by weight. For example, unauthorized copies of the current sun-ray style Maltese cross with eagle badge weigh 0.75 ounce as compared to 1.5 ounces for an official badge of two-piece construction. An official firefighter's Maltese cross weighs .8 ounces in the original white metal construction and 1.3 ounces in the current plated copper construction. The copy illustrated with the flat back is 1.1 ounces. However, identification by weight has become irrelevant since unauthorized copies are currently made from the same dies and with the same process as officially manufactured badges.

Another form of reproduction badge is a retiree badge. Until recently, as noted in Chapter 2, members were not allowed to retain their Departmental badge upon retirement unless they had completed thirty years of active duty. Badge manufacturers would, however, make a badge for a retired officer in the same style as an official badge except that within the number plate, the letter "R" or "RET" would replace the member's assigned number. An official badge was, for a very short time, issued to retired members starting in 1899 (see Chapter 13).

Retirement badges. These are not issued or authorized by the Department.

It appears to be relatively common that individuals with some connection to the Department have badges made on their own. Typically, these vary distinctly from official Departmental badges and are therefore easy to spot.

Individuals working with or for the FDNY who are not issued official badges often have unofficial ones made like this example from a security officer at FDNY headquarters in 1997.

Although the FDNY did have a canteen, badges were not issued by the Department to those who staffed it. This badge has a New York State seal which is never used on official FDNY badges.

As noted in Chapter 11, the official badge issued to Auxiliary Dispatchers during World War II were celluloid pin-type badges. Two unofficial badges bearing the title Auxiliary Dispatcher have been found. The first is a standard FDNY pentagon badge with the title "Aux. Dispatcher" in the upper banner. On the reverse is a round disk with a name engraved. This feature is uncommon and strongly suggests that this badge was not issued by the Department. A more contemporary sun-ray Maltese cross badge with the title "Aux. Dispatcher" was also found. The badge itself is discernibly different from the official FDNY sun-ray Maltese cross and is most likely struck from a different die, presumably a manufacturer's generic blank. The city seal in the central disk has blue enamel, as does the lettering on the badge, and is not in reverse. All of these features confirm that this is not an official badge.

Another badge with the title Dispatcher was found. It is a basic FDNY pentagon, however the city seal, as in the previously described badge, has a blue background. The pentagon badges did not have enamel backgrounds on the city seal and only one official FDNY badge has ever been made with blue enamel, the Theatre Detail badge.

In 1946, the Department introduced its first canteen vehicle to serve refreshments to members at the scene of multiple alarm fires. In 1958, the Department decided to retire the vehicle and not to replace it. Fortunately, out-

The FDNY Canteen.

Series of unofficial Dispatcher and Auxiliary Dispatcher badges.

side organizations including Buff clubs, the Salvation Army, and the Red Cross took over this function and continue to provide refreshments in this tradition to the current day (see Chapter 16).

A sun-ray Maltese cross badge with the word, "Canteen" was apparently made by or for a civilian involved with one of the latter canteen services. It is a gold-plated badge with "Fire Department - City of New York" in the circle around the New York State seal in the center. All enamel is blue which, as indicated earlier, is strong evidence that this is an unofficial piece. The use of a State, instead of City seal, and a flat back further confirms that this badge is not authentic.

In 1965, a member of the Marine Bureau commissioned the C. G. Braxmar Company to make a belt buckle resembling the original Marine Engineers and Pilots badges. They were struck from the original dies, with the fireboat *Firefighter*. Some members were known to replace the buckle findings with a screw post and to wear it as an unofficial cap badge. Some of these altered badges may have entered the collector's market but should be easily identifiable by the remnants of the buckle findings.

Belt buckle made from original Marine Pilot die.

Sterling silver reproduction of early 1900's fire lines badge. Made by Steve Scher in 2001.

Steven Scher, designer of the 125th Anniversary badge, introduced a reproduction of the "alarm box" fire lines badge in 2001 (see Chapter 13). Exact in every detail, purchasers were able to have their own names engraved on the badge. One departure from the original practice was that reproduction badges were available in a variety of metals including nickel, sterling silver, antique gold and solid gold. Aside from the uncommon metals, these reproductions can be easily distinguished from originals by the pin and clasp that are obviously not of an early 20th century vintage.

Items associated with the FDNY are highly sought after by collectors. So, it is no surprise to find some historic badges reproduced. What is surprising is that many "knock-off" badges are available, much the same as cheap designer watch copies on the streets of the City. A Taiwanese company was sued by the designer of the 125th Anniversary Commemorative badge (see Chapter 17) for manufacturing and selling a low-quality counterfeit. The counterfeits are discernable by the inexpensive, shiny, plating that is applied instead of the rich, bronze color of the original. Also, they were not sold in the presentation cases as were the originals.

This "knock-off" copy (right) of the 125th Anniversary commemorative badge was manufactured in Taiwan in 1991

It is common for members of the Department to have a replica made of their badge for a family member. These badges are commonly called "Courtesy Badges." Typically, they are identical in appearance to the member's badge except that they are about 75 percent smaller. Badges are also often reproduced in some form of jewelry, such as a charm or pin, for a spouse or significant-other, if not for the member him/herself.

Courtesy badge made by Braxmar, alongside a standard issue badge.

One interesting badge was owned by Edward Waterman (see Chapters 12 and 14 for Mr. Waterman's other badges). Mr. Waterman had an FDNY Photographer badge that was evidently unauthorized. At first glance it would appear to have been made from a casting since the back is flat and uneven. However, the number plate is also very different than FDNY badges, leading one to the conclusion that if it was cast, it was not from an FDNY badge. Also, the number "9" is applied instead of engraved on the plate. This badge is in the hands of Mr. Waterman's grandson and is therefore certain to have belonged to him. How it came to be issued to him is uncertain. It is possible that one of his many friends who were Department photographers, such as Raymond Hielriegel, may have procured it for him.

Unofficial Photographer badge (with applied rather than engraved number) made, by someone, for Edward Waterman.

Uniformed members are routinely assigned to perform specific duties, often on a permanent basis. On occasion, badges have been made for each duty. However, these badges are not authorized or issued by the Department. For example, uniformed members may be assigned to the Community Relations Bureau to perform such tasks as interacting with the media during significant fires or emergencies. In some cases, the members assigned to special duties have had custom badges produced. One example, shown here, is a standard FDNY sun-ray Maltese cross with the title, "Press Officer" in the ring and the radio identifier engraved in the number plate.

All Chief officers assigned to firefighting duty are assigned an Aide. The Aide serves as the Chief's driver and assists at the scene of a fire by transmitting and receiving radio communications to and from the Chief and performing other tasks. A distinctive helmet frontpiece is issued to identify Aides at the scene. An official badge is not issued for this position. Several versions of unoffi-

Badges utilized by uniformed FDNY members for specific assignments. These are not issued or authorized by the Department.

cial Chief's Aide badges have been found. Early renditions are based on the standard FDNY Maltese cross. More recently, a vendor has offered one in the sunburst style with a panel for engraving the specific assignment. In the example shown, the letters "QBFC" stand for Queens Borough Fire Command.

Other examples of assignment badges include those for Executive Officer of the Bureau of Health Services, Chief of Marine Bureau, and Badge and ID Desk. The Chief of Marine Bureau badge has blue enamel, another sure clue that this is not an authorized badge.

Unofficial badge manufactured and sold to general public as an FDNY/EMS badge.

In 1997, a vendor of collectible badges marketed a badge that was supposedly issued by the FDNY for the new Bureau of Emergency Medical Services. However, this was *before* a badge was designed for uniformed members of the Bureau. The badge is a Maltese cross. It has a profile similar to the old Long Island City (paid) F.D. badges, not the current FDNY fireman's badge. In the upper and lower arms are the words, "Fire" and "Rescue" respectively, a phrase not used by the FDNY in reference to BEMS. In the center is a fire scramble, and the words "City of New York EMT" are displayed in the concentric ring. Neither of these have ever been featured on an FDNY badge. This unauthorized badge was sold to the general public for $29.95.

One legitimate badge that many mistakenly believe is issued by the FDNY is the Stationary Fireman badge. It was issued by the Stationary Firemen and Engineers union. These individuals operate the thousands of boiler plants located throughout the city. Their jobs are roughly

Unofficial FDNY Auxiliary badges.

Badge officially issued by Local 1795 of the Stationary Fireman's union; an engineering not a fire fighting job.

equivalent to what the Department called Stokers. The title fireman has been used for many years by individuals who operate steam engines but, for obvious reasons, it could not be used by the men doing this job in the fire department. The badge is a sun-ray style Maltese cross without eagle. Engraved on the central ring is, "Stat Fireman - City of New York." It has a New York City seal with blue enamel background in the center. A panel on the upper arm reads, "Local 1795." The badge is silver in color and is no longer distributed.

As noted in Chapter 12, Honorary Firefighter badges are occasionally presented to individuals. However, these are not officially issued by the Department nor are the recipients recognized as honorary members. The badges are typically commissioned by individuals or companies and given to Buffs or others as tokens of gratitude. Two examples of these unauthorized badges are shown.

Badges probably made by members of the Department, bestowing unofficial honorary status.

One interesting badge raises a perplexing question; is it authentic? It was made for a member of the Department's headquarters staff. The badge is in the form of a wreath with a plate in the middle displaying the letters, "F.D.N.Y." It was intended to be worn on a warm winter hat. But a similar hat badge can be seen in a photograph of a group of uniformed firefighters on an

A unique badge made by a Department official to place on his unofficial winter hat. This badge is one-of-a-kind.

Ahrens-Fox photo of a new FDNY pumper taken in front of the quarters of the Cincinnati Fire Department. The Cincinnati fireman's cap device could easily be mistaken for an FDNY badge without the proper knowledge.

Two badges appearing to be made from official dies. On the left, is a Captain's pentagon; it should be silver-tone. Above, no such badge was ever officially issued for Chief Marine Engineer.

One-of-a-kind prototype badges.

An unusual flat badge, struck with a die for the FDNY Maltese cross but not struck using the "force" from the rear. The number "24" is permanently applied.

Drawings from the archives of C. G. Braxmar Company with conceptual badges proposed for the FDNY.

FDNY Ahrens-Fox engine. The photograph was taken at a Cincinnati fire station near the Ahrens-Fox factory. The firefighters belong to the Cincinnati Fire Department. Without knowing the history of both this badge and the photograph, one could certainly be led to the incorrect assumption that the FDNY issued a hat badge in the form of a wreath.

It is not always possible to determine the origin of badges that have some kind of designation suggesting a connection to the FDNY. In some cases it is presumed that these one-of-a-kind pieces are prototypes. Badges with no number or the number "1" are likely prototypes or phonies.

Throughout the history of the FDNY, badges have been conceived and proposed by the Department or by designers and manufacturers. In some cases, prototype badges were first created for consideration and review.

A series of historic and very detailed artist's drawings are in the archives of the C. G. Braxmar Company. They show a variety of badge designs, most of which were never actually made. Some of these drawings are shown here.

CHAPTER 19

Presentation Badges And Badge Jewelry

Presentation Badges

Badges are often given as gifts of appreciation to members and non-members. Typically, presentation badges are given by groups or individuals, such as line organizations or companies, not by the Department. They are often anniversary or retirement gifts.

Presentation badges are custom and personal. The number of presentation badges in existence is not known but is presumed to be large. The illustrations in this chapter represent only an handful of these badges.

Since presentation badges are gifts and not really intended to be worn, they are especially attractive. They are routinely made of precious metal, typically solid gold, and often have diamonds or rubies. Most of the presentation badges have an inscription on the back. During the early part of the twentieth century, firefighters were presented with badges by the members of their company on such occasions as an anniversary or retirement. Badges from this era in both Maltese cross and pentagon shapes can be found. Only one badge from the post-1937 era, in the sun-ray Maltese cross design, was located but is not shown here. Most presentation badges are jeweler-made, but some were made by badge manufacturers. C. G. Braxmar probably produced the greatest number and had an impressive assortment of presentation badges in their catalogs of the period.

The oldest, and perhaps most unusual presentation badge found is made of ivory. It is shaped like a helmet frontpiece, and is engraved like a helmet front with the company name and number "Union 18" and the member's initials "PMcC." It has no pin or other attachment device on the reverse. However, it is grooved and may have been affixed to some other item, such as a walking stick or cane. It is believed that this badge was presented to "Paddy" McCaffrey of Union Engine 18 who, along with four drivers from the 42nd Street trolley line, rescued twenty children from a fire at the Colored Orphans Asylum during the infamous draft riots in July 1863.

Perhaps the oldest presentation badge known made of ivory. Presented to Patrick McCaffrey of Engine Company 18, credited with saving children from the burning Colored Orphan's Asylum during the Civil War draft riots in 1863.

Fireman Paddy McCaffrey of Union Engine 18 rescued 20 children from the Colored Orphan's Asylum during the draft riots of 1863.

In the early 1900's, the *New York Daily News* held a contest to select a "Captain of Captains," one representing Manhattan and the Bronx and one representing Brooklyn and Queens. They were each to be awarded a badge described in an article in the *Daily News*:

> *"Each badge will be of solid 18-karat gold, surrounded by sixty fine diamonds, graduated in size. The word "CAPTAIN," made of fine rose cut diamonds, will be raised on a ribbon of dark blue enamel. The rest of the ribbon will be enameled turquoise blue. The words "FIRE DEP'T. NEW YORK," will be cut into the band and enameled in dark blue. The Coat of Arms will be modeled in relief on transparent, dark blue enamel, and set with five fine diamonds. The Coat of Arms shield will be surrounded by a circle of forty selected fiery rubies."*

Exact Description of the Thousand-Dollar Badges

Gold and diamonds and rubies, and fine enamel will form the two jeweled badges which the NEW YORK DAILY NEWS will present to the winners of the great contest conducted under its auspices to determine the two most popular captains in the paid Fire Department of Greater New York. One badge will be given to the most popular captain in the Boroughs of Manhattan and the Bronx; the other will be given to the most popular captain in the Boroughs of Brooklyn and Queens.

Each badge will cost $1,000, and will be exactly alike and will conform closely to regulations, as shown in the accompanying cut, made from the design furnished by Lambert Bros., jewelers, of Third avenue and Fifty-eighth street, who made the diamond shields awarded by this paper to the two most popular police captains.

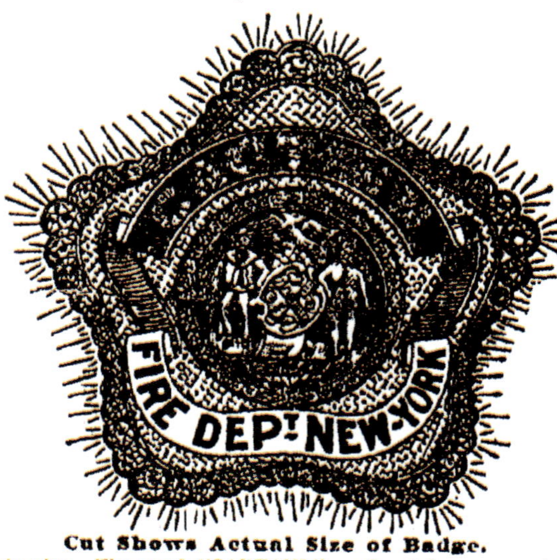

Cut Shows Actual Size of Badge.

Each badge will be of solid 18-karat gold, surrounded by sixty fine diamonds, graduated in size. The word "CAPTAIN," made of fine rose cut diamonds, will be raised on a ribbon of dark blue enamel. The rest of the ribbon will be enameled turquoise blue. The words "FIRE DEP'T, NEW YORK," will be cut into the band and enameled in dark blue. The Coat of Arms will be modeled in relief on transparent, dark blue enamel, and set with five fine diamonds. The Coat of Arms shield will be surrounded by a circle of forty selected, fiery rubies.

As emblematic works of art the badges will be unsurpassed. There will be the gold to typify devotion to duty; the hard enamel to typify the arduous work and also the familiar color of the fireman's uniform the diamonds of the first water to typify the flame-subduing element, and the blazing rubies to typify the flames themselves.

Article from the New York Daily News *about the "Thousand-Dollar Badges" to be issued to the "Captain of Captains."*

"Presented to John J. McElligott, Chief Fire Dept, by the Officers and Members Naer Tormid Society, Inc. of the Fire Department, March 1, 1932"

"Presented to Chief John J. McElligott by the New York Firemen's Cycle Club promoted to Chief of Department Fire Department New York, March 1st, 1932"

"Presented to Hon. William Einziger by his loyal friend of Staten Island Hon. Thomas J. Drennan Fire Commissioner City of New York, July 6, 1922"

"Presented by Officers and Members of Engine Co. 275 to Lieutenant Frederick J. Lepold, Nov. 15, 1922"

FIREMEN GET GOLD BADGES FROM COMRADES

Battalion Chief Henry P. Kirk. Lieut. Francis H. McLaughlin. Fireman John Bedell.

A complete surprise was sprung on Battalion Chief Henry P. Kirk, Lieutenant Francis H. McLoughlin and Fireman John H. Bedell, who have just completed twenty years service in the Fire Department, in the headquarters of Engine Company No. 207, on Pearl street, last night, when they were presented with diamond badges. Several hundred friends of the firefighters were present to see them signally honored by their comrades.

For the past few weeks a committee of seven, consisting of Captain Frank L. Owens and Edward B. Donlon, No. 207; Thomas O'Toole and George Hartigan, No. 208; John Kelly, No. 205; John McNamara, No. 206, and David Murphy, Hook and Lader No. 103, have been doing missionary work among the members of their companies to purchase a diamond badge for their chief, while the members of No. 207 were getting together to present a similar token to Lieutenant McLoughlin and Private Bedell. While the preparations were going on the recipients of honor were kept in complete ignorance of the full plans.

Yesterday was the twentieth anniversary of their service in the Fire Department, and it happened that it was the chief's day off. McLoughlin was on his vacation, while Bedell was on duty on the main floor. Chief Lally of Brooklyn and Queens instructed Chief Kirk and Lieutenant McLoughlin to report at No. 207 firehouse at 9 o'clock. The crowd began to gather around 8 o'clock, and Bedell, at the door, was thinking what a fine time they would have upstairs celebrating the chief's anniversary, forgetting the fact that he had just rounded out twenty years himself.

When the chief and lieutenant appeared on the scene they were ushered to the top floor of the house, and Bedell was also summoned upstairs by Acting Chief John Fitzgerald. When the trio entered and beheld the large gathering consternation and surprise was plainly visible in their faces. They were escorted to an improvised platform, where they took seats with Chief Lally. When the applause had subsided, Chief Lally stated the reason for his ordering his subordinates to report. It was the first time in his life that Chief Lally had ever consented to make a presentation speech to any of his men, although often importuned to do so, but in view of their records he said it was a pleasure to act as spokesman for the men of the battalion. His remarks were rather brief. He called attention to the work of the guests, commented upon their excellent habits, and said it was an honor for him to be with them on the occasion of their twentieth anniversary.

Of course, the trio were called upon to make speeches, but they were so completely surprised that words failed them. However, all took occasion to thank their comrades. After the ceremonies the gathering enjoyed an excellent vaudeville entertainment. John Casey, Joseph Lind, Robert Jones, W. J. O'Connell, Joseph McKenna, Edward Montague, Jack O'Donnell and Joseph Collins amused them with songs and recitations, while the committee under the guidance of Edward Donlon saw to it that they were well supplied with cigars and refreshments.

There is no gainsaying the fact that the trio honored last night have excellent records. Chief Kirk is the only fireman in the borough today who ever secured promotion for bravery. Some years ago, when the Arbuckle Building was on fire, Kirk, who had just been in the service but a year, climbed an 85-foot ladder and then mounted another to one of the top floors of the burning structure five times and rescued five people. There were no medals awarded in those days in Brooklyn and the only recognition that he received for his perilous deed was an appointment as an assistant foreman. Since that time he has steadily been promoted, and in 1905 he was made the chief of the Thirty-first Battalion, which includes four companies and a hook and ladder, and which covers the most important district in the city—the downtown section.

For ten years Lieutenant McLoughlin has been attached to companies in the heart of Brooklyn, and in more recent years he has been attached to Chief Lally's staff. In this capacity he has attended many of the large conflagrations that happened in the city, but has never had the opportunity of distinguishing himself so signally as his chief.

Fireman Bedell also displayed his metal some years ago while trying to stop a runaway. A horse was running wild down Gold street, and was headed toward five little children on the sidewalk. Bedell ran toward the animal and bore him down, but with severe injuries to himself. The beast trampled all over his legs and thighs and Bedel was laid up in a hospital for ten weeks.

See this page tomorrow for the continuation of E. B. Lent's story of the progress of Long Island. It deals with Southampton and Easthampton.

Brooklyn Eagle newspaper article, September 12, 1913.

"Presented to Lieutenant Francis H. McLaughlin by the Officers and Members of Engine Co. 207, Sept. 11, 1913"

The badges were created by the Lambert Brothers jewelers of New York City at a cost of $1,000 each. The significance given to this design, which was in the same shape and size specifications as the uniform pentagon badge, was described as follows:

> *There will be the gold to typify the devotion to duty; the hard enamel to typify the arduous work and also the familiar color of the fireman's uniform; the diamonds of the first water, to typify the flame-subduing element; the blazing rubies to typify the flames themselves.*

The *Daily News* continues to recognize the contributions of New York's firefighters and police officers with a monthly "Hero Award" consisting of a certificate and cash prize.

"Presented to Irvin P. Folley by the Officers and Members Engine Company 240, 1899-Nov. 16-1919"

"Presented to Captain Henry M. Keil by Officers & Members Engine Co. 25, July 9, 1933; 1913-1933"

"Presented to Patrick Walsh by Officers & Members of 1st Battalion, December 10, 1921"

"Presented to Dr. John J. White by the Medical Officers of the New York Fire Department, May 9, 1927; 1907-1927"

"Presented to Edward Farrell by the Officers and Members of the Fire Alarm Bureau, July 9, 1922; 1902-1922"

"Presented to W. H. Snyder by the members of the Fire Alarm Telegraph Bureau Upon the Completion of his 25th year of service."

Small solid gold presentation badge to William Willis in 1908.

"Presented to Stephen F. Sullivan by the officers and members of H&L Co. 117 F.D.N.Y. as a token of esteem, Nov. 21, 1897 Nov. 24, 1917"

"Presented to William A. Aldrich by Officers & Members Engine Co. 253 NYFD; 1896-Feb. 1 - 1916"

"Presented to Fireman Henry McGinness by the Officers & Members of Engine Co. 261, July 7th 1916"

"Presented by the Officers & Members of H&L Co. No. 105, Sept. 11, 1913"

"Presented to Foreman Francis M. Fuchs by the Officers and Members of Engine Co. 203, Dec. 20th, 1906"

"Presented to Fireman David Stark by the Officers & Members of Engine Co. 4, July 10, 1927;1907-1927"

"Presented to Lt. Patrick J. O'Brien by the Officers and Member of Hook & Ladder Co. 22, October 1, 1940; 1920-1940"

"Presented to Fireman Carl A. Montelin by the Officers and Members of Engine Co. 279; April 6, 1926; 1906-1926"

"Presented to Fireman Kenneth E. Fender by the Officers and Members of H&L Co. 126 and Eng. Co. 303 and 13th Div.; 4-1-1950"

"Presented to Frank Ulrich by the Officers & Members of Engine Co. 222 F.D.N.Y. June 11th 1921, 1901-1921"

A very unusual presentation badge with the title "Official Agent" without an inscription on the reverse. Based on the hallmark, this was issued between 1900-1910.

A rare presentation badge for 33 years service to the FDNY.

Pocket watch "Presented to James McCutchen by the Officers and Members of Engine Co. 43 NYFD for 33 years of active service, Sept 18, 1865-Sept 18, 1898"

"Presented to Matron Sarah Cope by the Officers and Members Engine Co. 21 F.D.N.Y. as a token of esteem, 1890-Sept 12-1910"

Mrs. Ellen Dunn celebrating 50 years as a Matron with the FDNY in 1944.

Gold-plated 20-year presentation badge without jewels or inscription. Made by the C. G. Braxmar Co.

A rare Fire Patrol presentation piece inscribed "Presented to Lieutenant Louis C. Ammermuller by the Officers and Members of Fire Patrol No. 2 1893 - Oct 1"

Chief Valentine Bangert of the Jamaica F.D. wearing a presentation badge circa 1907.

"Presented to Cornelius E. Cain by his company, September 1908"

"Presented to William J. Glazebrook by the South Side Engine & Pump Co. of Woodhaven L.I."

"Presented to William Leitz by the Officers and Members of Engine Co. #9 NFD, August 5th 1913"

"Presented by Port Richmond Engine Co. for best fire record, April 21, 1891"

"Presented to T. F. Cruise by his company; December 1905"

Badge presented by the Firemen's Association of Greater New York on the occasion of twenty years of service to the FDNY.

Gold FDNY Maltese cross charm for bracelet given by fireman to his wife, circa 1960.

Pocket watch and fob with the following inscriptions: (Watch) "Presented to Lieutenant Frank Brownley by the Officers and Members of Engine Co. 285 and Hook & Ladder Co. 142, May 5, 1925; 1905-1925" (Fob) "Presented to Frank Brownley by the Officers and Members of Engine Co. 251 NYFD, May 1, 1918"

Badge Jewelry

FDNY badges can be found in many different forms of jewelry. They include rings, lapel pins, and charms for necklaces or bracelets. Most are jeweler-made. In the latter part of the twentieth century, the 20-year ring appears to have replaced the 20-year presentation badges.

It was also common practice to present gift helmets. These helmets have a presentation plate in the shape of a pinched-shield badge on the rear brim. When detached from the helmet, these items are often mistaken for badges.

A small selection of badge jewelry is shown here.

"1904 - 1924, Presented to John L. Roth by the Officers and Members of Engine Co. #7, April 23, 1924"

Pocket watch with inscription: "Presented to Gus Kettler, Officers & Members of Rescue Co. No. 2, Oct. 28, 1931; 1911-1931"

Watch fob presented to John McElligott upon his appointment as Battalion Chief, with the inscription: "Presented to Captain John J. McElligott by the Officers and Members of Hook & Ladder Co. 8, April 21, 1914"

Watch fob inscribed, "Presented to Charles Sprinka by the Officers and Members of H&L Co. 117, April 4, 1933; 1913-1933"

Sterling silver "sweetheart" pin in the shape of an FDNY Maltese cross with the member's badge number.

Generic FDNY Maltese cross ring with red stone (circa 1960).

Watch fob with the initials "E.A.S." Engine 33.

Lapel pin badge made by or for a member of the Emergency Auxiliary Corps assigned to Engine 320, circa 1941.

Presented to H.C. Jones by Hollis H&L Co. No. 1, August 1921.

Twenty-year service badge issued to members in the FDNY who began their career with the Brooklyn Fire Dept, prior to the consolidation in 1898.

Money clip with replica of a Third Alarm Association badge.

Small lapel pin badge presented to E. A. Waterman by members of FDNY Engine 15.

Very small lapel badge dated 9-1-1939. This may be a date of appointment or retirement.

20-year ring inscribed: "Ray Urbanowicz from members of the Medical Division 20 years 6-16-73"

CHAPTER 20

Bureau Of Emergency Medical Services

On March 17, 1996, the New York City Emergency Medical Service (NYCEMS) was consolidated into the FDNY, increasing its size by more than 3,000 personnel and 650 vehicles. The interesting history of ambulance services continues today under the auspices of the FDNY.

The first public ambulance was established in New York in 1869. It was based at Bellevue Hospital, the principal municipal hospital of New York City. This first ambulance was horse drawn and staffed by a physician and driver. By 1878, two private facilities, New York Hospital and Roosevelt Hospital, also provided ambulance services. The first motorized ambulance to be introduced in the City was at St. Vincent's Hospital in 1900. During this era, all ambulances were hospital-based

NYC Department of Hospitals Ambulance Engineman and Surgeon at Bellevue Hospital, circa 1948.

Cap and breast badges from a NYC Department of Hospitals Engineman, or driver.

1960's Foreman's badge from the NYC Department of Hospitals. Foremen were supervisors who might be given the rank of Captain today.

and there was no centralized coordination. Anyone needing an ambulance would contact the hospital directly. Distinctions between ambulances based at municipal and private hospitals were relatively insignificant. In 1929, the New York City Department of Hospitals was formed and along with more modern technology, coordination and centralized dispatch of municipal ambulances began.

On May 29, 1969 state legislation was introduced for the organization of the New York City Health and Hospitals Corporation (HHC) to succeed the Department of Hospitals. This public benefit, non-profit entity was charged with overseeing the country's largest municipal health care system. On July 1, 1970 the New York City Emergency Medical Service (NYCEMS) was created in accordance with the legislation, under the control of the HHC. NYCEMS became the City's third uniformed service. This model is found in several large metropolitan areas such as Boston. In most cities where ambulance service is provided by a municipality it is part of the fire department, as in Chicago, Philadelphia, and Seattle. During the relatively brief period when NYCEMS was an independent entity, the subject of consolidating it into the FDNY was often raised, particularly when the media criticized ambulance response times. Despite these discussions, NYCEMS's independence strengthened and in 1977 it was designated the coordinating agency for all ambulance services in the City including municipal, private hospital, and volunteer.

The FDNY has had ambulances in its fleet since 1914. However, their use was limited to FDNY personnel only. Although they responded to greater alarms for the purpose of providing emergency care and transportation for injured members, their role was primarily for non-emergency, off-duty transport. The Medical Division, later the Bureau of Health Services, has had a long history of providing emergency care at the scene of greater alarms with ambulances as well as a fleet of mobile emergency rooms staffed with Department Medical Officers. The provision of emergency care to firefighters at the scene by Dr. Archer is legendary and a unique part of FDNY history. The growth of NYCEMS from 1970 through the mid-1990's superseded this Departmental function. Now, with its full integration, emergency medical care will once again be kept in the FDNY family.

The first FDNY ambulance in 1914 was used only to transport injured firefighters.
(Photo courtesy of John A. Calderone.)

FDNY Mobile Emergency Room.
(Photo courtesy of John M. Calderone.)

FDNY ambulance introduced with the merger of NYCEMS and the Department in 1996. (Photo courtesy of John M. Calderone.)

For the most part, the Department's role in emergency medical services to the general public has been incidental to responding to calls that are non-fire emergencies. However, several notable EMS initiatives have been undertaken. In October 1872, signals for transmitting ambulance calls over the fire alarm telegraph system were initiated. In September 1975, the Department introduced a training program for firefighters to become state-certified Emergency Medical Technicians. For the first time, fire apparatus were equipped with first aid supplies. Firefighters were so enthused about this early program that over 1,200 applied for the 320 available slots in the first training program. In 1995, a Certified First Responder - Defibrillation program was instituted whereby select engine companies were equipped with defibrillators and trained firefighters to operate them. FDNY was clearly positioned to accept the responsibility of providing comprehensive emergency medical care in 1996, from first response through basic and advanced life support, throughout the five boroughs.

Assimilating the Bureau of Emergency Medical Service (BEMS) necessitated numerous operational changes in service and administration. In 1870, the first New York City municipal ambulance responded to 1,401 calls. In 1997, the first full year of consolidation, BEMS responded to approximately 1.3 million calls.

NYCEMS replaced its green uniforms with FDNY blue. Since Emergency Medical Technicians, Paramedics, and BEMS officers are uniformed but are not firefighters, it was necessary to develop a system of identification consistent with the FDNY but distinct from firefighting personnel. Duty uniforms and turnout gear were designed and issued. Large white lettering on the back of these uniforms read, "FDNY EMT" "FDNY PARAMEDIC" or rank-specific lettering for officers.

Dress uniforms worn by firefighting personnel were issued to BEMS as well, but with a few distinctions. First, the traditional bell hats worn by officers are blue instead of white. Second, only cap devices were authorized. No breast badges have been designed or issued.

Upon consolidation of NYCEMS into the FDNY, Commissioner Von Essen ordered that BEMS personnel were not to be issued or to wear badges while on duty, although he allowed them to retain their NYCEMS shields for identification purposes. Former NYCEMS cap devices displayed the New York City seal, as worn by NYPD and other uniformed City agencies. The new cap devices are round with a blue and white enamel Star-of-Life symbol in the center. The enamel is set in a metal frame that is either silver or gold-tone, depending on rank.

Dress cap of an FDNY BEMS Chief.

For historical purposes, pre-FDNY NYCEMS badges are included in this book. During the days of the Department of Hospitals, ambulance personnel badges were similar in appearance to the NYPD pinched shield design. During HHC administration, a unique eight-sided badge was introduced. Officers' badges are similar in design to NYPD sunburst shields, including leaves and stars for Chief Officers. In the late 1980's, the octagon badges were replaced once again with NYPD-style shields.

Although there are no official BEMS badges, someone manufactured a badge and sold it to novice badge collectors as a BEMS badge. This false badge is illustrated in Chapter 18.

Badge issued upon the formation of the New York City Emergency Medical Service in 1970. It was replaced by the octagon-shaped badges in 1978.

Chief Corpsman's badge circa 1970-1977.

Octagon-shaped badge issued to Emergency Medical Technicians and Paramedics from 1978 through the late 1980's.

Officers' badges used by NYCEMS from 1978 through the late 1980's.

Officers' badges used by NYCEMS from the late 1980's until consolidation with the FDNY.

Last badge issued by NYCEMS prior to its merger into the FDNY.

CHAPTER 21

Federal, State, City & Private Fire Brigades

There are, or have been, several firefighting brigades that are not part of the FDNY but are essential to comprehensive fire protection in the City of New York. The history of these brigades is discussed in this chapter, even if there is no documented badge so that collectors who happen upon badges with a New York City designation might research their possible origin.

Private Fire Brigades and Departments

1939 and 1964 New York World's Fairs

New York City has been the host of three world expositions, beginning with the first held in the United States in 1853 at the infamous Crystal Palace. The Crystal Palace was destroyed by a landmark fire in 1858. At the World Columbian Exposition in Chicago in 1893, twelve firemen and eight workmen were killed fighting a fire in the Fair's icehouse. As a result of these fires, special fire protection was planned for the 1939 New York World's Fair.

Initially, the New York World's Fair Corporation planned to establish its own fire department at the fairgrounds in Flushing Meadow Park, which was previously the Corona Dump. They hired Thomas F. Dougherty to lead this department. Dougherty, a member of the FDNY from 1888 through 1932, was a former Assistant Chief, a Bennett Medal winner, and the namesake of an FDNY valor medal. The Corporation purchased five pumpers from the Ahrens-Fox company. At some point, the New York World's Fair Corporation decided not to man the equipment with its own firefighters. In September 1938, the FDNY 55th Battalion was formed and stationed at the Fair to man the apparatus. It included three Battalion Chiefs, fourteen company officers, and seventy firemen manning one double and two single engine companies. In addition, sixty-five "Fire Guards" were organized under the direction of Chief Dougherty. The Fire Guard was a private force of retired FDNY members hired by the New York World's Fair Corporation. They operated in a fire prevention capacity, inspecting buildings and patrolling the grounds during the hours that the Fair was open to the public. The Battalion and the Fire Guard were disbanded in 1940 shortly after the close of the Fair. The apparatus was given to the FDNY.

The only badge located from this event was Chief Dougherty's. It is an eagle-topped starburst design with a reverse enamel center of blue depicting the "Trylon and Perisphere" symbols of the Fair. It has a solid back, bears the date "1940" and the wording, "World's Fair Fire Department." This is a working badge, not a presentation badge, with double loop attachments for a bar pin.

Badge of Thomas F. Dougherty, Chief of the 1939-1940 New York World's Fair Fire Department.

The Department did not provide any on-site coverage at the 1964-1965 New York World's Fair. Pinkerton's International Detective Agency was contracted to provide a number of essential services at the Fair, including fire protection. (Pinkerton's also provided security services at the 1939 NYWF.) The World's Fair Fire Department (WFFD) was staffed by 100 retired FDNY members, commanded by former Assistant Chief Thomas P. O'Brien. A manual written for the department was clearly modeled after the FDNY Rules and Regulations, though much abbreviated. Three specially designed pumpers, painted in the Fair's blue and orange colors, were each manned by one Captain, three Lieutenants, and twenty firemen on a platoon system deployed in three fire stations. (A vehicle badge from one of these units is shown in Chapter 17.) In addition, the WFFD provided a chauffeur for each of the four ambulances of the World's Fair Medical Department. Chief O'Brien also had one Assistant Chief, retired FDNY Battalion Chief James E. Gowdy. The unit was disbanded at the Fair's conclusion in 1965.

Badges issued to the 1964 WFFD utilized a Maltese cross. Engraved in a circular ring are the words, "World's Fair Fire Department." Captains' and Lieutenants' badges were a sun-ray Maltese Cross without eagle with their rank engraved on a top panel. Assistant Chief Gowdy's badge was of a standard style commonly recognized for its issuance to NYPD higher-officers. It bears the title, "Assistant Chief" in reverse blue enamel. Chief O'Brien's badge was not located while researching this book.

In the center of all WFFD badges is a reverse blue and orange enamel disk behind the Unisphere symbol of the Fair. The New York World's Fair Corporation insisted that all items marked with the Unisphere bear an appropriate trademark label. Therefore, all WFFD badges have this trademark on their reverse side. Any badge *not* bearing the trademark can be assumed to be a reproduction.

FIRE SWEEPS WORLD'S FAIR STORAGE BUILDING - NEW YORK CITY — Firemen of the World's Fair Fire Department, aided by smoke-eaters of the New York City Fire Department, are shown as they brought a hose into play in combating a fire which swept a one-story storage building on the site of the New York World's Fair 1939, July 8th. They succeeded in extinguishing the blaze after it had destroyed reserve furniture and stores of paper. [Original Photo Caption]

Men of the 1964 New York World's Fair Fire Department in front of the Fair's symbol, The Unisphere. (Photo courtesy of Pinkerton's, Inc.)

Badges from the 1964-1965 New York World's Fair Fire Department.

United Nations Safety & Security Service

In 1950, the United Nations moved into a seventeen acre complex along the East River. The predominant structures include the thirty-nine story Secretariat Building and the 316,550 square foot General Assembly Hall. The entire complex is officially not land of the United States, nor that of the City or State of New York. It maintains its own postal service, police force, and fire department. The latter two duties are performed by the United Nations Safety & Security Service.

Until 1995, the Service had personnel who were exclusively assigned to fire duty. However, since then all personnel are part of the armed security force with twelve serving on fire detail. The detail is trained and certified as an industrial fire brigade. Although this authorizes the Service to fight fires, its primary role is evacuation and containment. The FDNY is notified of any fires.

New York World's Fair Corporation logo on back of badge.

Obsolete badge for Fire Officers of the United Nations Safety & Security Service.

Distinct badges were issued to United Nations fire brigade personnel until 1995. They were identical to the badges worn by security staff except the word "FIRE" replaces the word "SECURITY." Ranks are placed in the lower panel. Today, security officers qualified for fire duty continue to wear an arm patch designating this qualification as a fire brigade member.

Corporate Fire Brigades and Fire Safety Directors

Since its settlement by the Dutch in 1625, New York has been a center of world commerce and is universally recognized as the home of leading financial markets. The headquarters of many companies are located in the City, some owning one or more of the skyscrapers that make up its world-famous skyline. As the City grew upward as well as in population, stringent fire codes were developed to avoid potentially catastrophic fires in high-rise buildings. Anyone who monitors the activity of the FDNY knows that specialized techniques and equipment have evolved over time to deal with the fire fighting challenges created by high-rise buildings. Encouraged by insurance companies, many corporations assisted in this endeavor by developing their own fire safety programs with the assistance and cooperation of the Department.

It is impossible to know how many of these early brigades existed, but it appears that many could be found throughout the New York area. The earliest known private fire brigade in New York was established on August 2, 1824 for protection in the dry dock area of Manhattan by the shipyards operating there. The company became part of the NYFD in 1828 as Live Oak Engine Company 44.

A 1905 report by the National Board of Fire Underwriters - Committee of Twenty named a few private fire brigades in Manhattan and the Bronx (see Chapter 6). These included R. H. Macy Company, John Wanamaker, New York Theater, Proctor's Theater, Bellevue Hospital, St. Luke's Hospital, New York Central & Hudson River Railroad, and the New York, New Haven & Hartford Railroad Company. The report states that private fire brigades existed in many department stores, hotels, theaters, hospitals, gas plants, steamship piers and railroad companies. The report indicates that the Committee of Twenty inspected sixty such brigades.

Staten Island seems to have been home to many companies with fire brigades long before its incorporation into the Greater City. These included the Barrett Brothers Die Works, S. S. White Dental Works, Proctor and Gamble, Bethlehem Steel-New York Shipyard and the New Brighton Die Works, which later became the U.S. Gypsum Company. These companies eventually became part of fully organized community departments. Barrett Brothers became Cataract Engine 2 of the North Shore Fire Department, S.S. White became Amicita H&L 1 of the Pleasant Plains Princes Bay Fire Department, and New Brighton became New Brighton Engine 4 of the depart-

Early corporate fire brigade badges.

ment by the same name. Private brigades or departments also operated at the Sea View Hospital and Sailor's Snug Harbor. Organizations in other boroughs also had brigades such as Hadden Chemical Company and the L&G Company in Brooklyn, and Johnson's Iron Foundry in the Bronx. The L&G Company, whose badge is shown, manufactured enameled tin products including early license plates and cookware. The Johnson's fire brigade badge is from the foundry of the same name. The Johnson's Iron Foundry was located on Spuyten Duyvil creek during the 19th century. The foundry made munitions during the Civil War and the Spanish American War.

Members of the International Ladies Garment Worker's Union Fire Warden program march in a parade along with the FDNY. The ILGWU lobbied for safer conditions after the infamous Triangle Shirtwaist Fire in 1911.

Landmark fires were often the impetus for creating programs to prevent and respond to fires. Within the corporate setting, the fire that stands out in any fire historian's mind is the 1911 Triangle Shirtwaist Company fire. One hundred forty-six women died in this blaze caused by highly combustible materials, blocked exits, and crowded working conditions. Thereafter, companies sought to prevent such disasters by forming independent fire brigades made up of volunteer employees.

Contemporary badges for corporate Fire Safety Directors.

In 1959, the FDNY became actively involved in establishing private fire brigades in commercial buildings within the City. A Fire Warden program was developed to provide standardized training to companies throughout the City. The training emphasis was on fire prevention. It included mandatory inspections conducted by the trained Fire Wardens who submitted reports to the FDNY Bureau of Fire Prevention. Graduates of the program were issued laminated identification cards by the Department. Badges were issued solely at the discretion and initiative of individual companies. Several private fire brigade badges are shown here.

Local Law Number 5, enacted in 1973, made New York the first city to require a certified Fire Safety Director in all buildings greater than 100 feet tall, or occupied by more than 100 people above or below street level or occupied by more than 500 people above the first floor. Several institutions provide training for the certification examination which is administered by the FDNY.

The Fire Safety Directors Association of Greater New York was formed in 1978 to promote and create a fire safe environment in high rise structures throughout the metropolitan area and to assist those concerned with the art and science of fire protection engineering. At the present time, it has 480 members representing such companies as Citibank, F. W. Woolworth, the Metropolitan Museum of Art, and New York University. Many of its members are retired from the FDNY. The badge shown for "Fire Safety Director - City of New York" is an unofficial badge which, like all of the others for Corporate Fire Safety Directors, is made by the individual or corporation.

The twin towers of the World Trade Center stood as proud and prominent components of the New York City skyline until they were destroyed on September 11, 2001 in the worst terrorist attack to hit the United States. The complex actually had six buildings in addition to the twin, 110-story towers. The first buildings opened in 1970 and the remainder were completed in 1977. The buildings occupied twelve million square feet of space on sixteen acres. Each floor of the towers encompassed 40,000 square feet, equivalent to one acre. Approximately 70,000 people worked in the World Trade Center and it hosted an estimated 50,000 visitors every day. It was owned by the Port Authority of New York and New Jersey which was headquartered there.

A dedicated Fire Unit of full-time Port Authority employees and twenty-five contracted Deputies were on duty around the clock at the World Trade Center. The Fire Unit was responsible for fire protection, life safety, emergency preparedness, code compliance, fire safety training and education, among other things. In addition, employees in the building, not necessarily employees of the Port Authority, were designated as Floor Wardens of the Fire Safety Team. Their role was to ensure that an alarm would be transmitted and that all personnel on their floor were evacuated if necessary. It may have been the effectiveness of the Fire Unit and the Floor Wardens that was responsible for so many people surviving the September 11th attack. The complex also maintained an on-site fire brigade composed of Port Authority Police Officers and a multi-discipline hazardous materials response team. Some of these officers were among the rescuers who perished on September 11th.

Fire at SOCONY-Vaccum plant in Greenpoint, September 13, 1919.

These celluloid badges were used by the Fire Safety Team at the World Trade Center during the 1970's and 1980's.

Badges were not issued to the Fire Unit and no special identification, other than turn-out gear, was used by the Port Authority Police fire brigade. Floor Wardens were issued a celluloid badge for identification. Two styles were issued, as shown. These were later replaced with a baseball-style cap which was much more visible than the badges.

The Port Authority continues to maintain fire services at several of its facilities, including New York's very active airports, bridges and tunnels.

The Standard Oil Company was one of many companies that operated oil refineries on the waterfront in Brooklyn, Long Island City, and Staten Island. After its forced breakup by the Supreme Court in 1911, the New York operation became known as Socony (Standard Oil Company of New York,) the predecessor of today's Mobil Oil. The company maintained a Mack fire truck at its Sone & Fleming Oil Works in Greenpoint. It was equipped with 250 feet of two-inch hose and three portable foam generators capable of producing 6,400 gallons of foam per minute. In 1953, the truck was supplemented with one built by the National Foam Company which had a 7,500 gallons per minute pump, 400 feet of two-inch and 800 feet of one-inch hose.

The company had a truck driver available twenty-four hours a day but its operation would be taken over by the FDNY during a fire. Several serious fires did occur at Socony's New York plants including the Sone & Fleming refinery in 1919, at the Long Island City refinery in 1907 and at the Staten Island bulk oil storage facility in 1960. No badges from these sites have been located, but badges

Brooklyn Union Gas Company Foam Apparatus Truck

from one of the company's New Jersey sites are known to exist. Badges for the company police force at the Sone & Fleming plant are in the Mobil Oil Company archives. The badge is made from a stock die with an eagle on top and a winged Pegasus in the center encircled by the company name. Fire badges, if issued, may have looked similar.

The Brooklyn Union Gas Company was founded in 1895 to provide natural gas for light and heat to the residents of the City. It continues to provide service to residential and commercial customers throughout the New York City and Long Island metropolitan areas. A fire brigade protects the company's liquid natural gas plant at its Maspeth facility. There, liquid natural gas is stored at a temperature of -259° F in order to supply gas to millions of homes and commercial buildings through a labyrinth of pipes under the City's streets. In addition to providing first-response to on-site incidents at the plant, the company's Director of Fire Safety serves as a critical link between the company and other firefighting agencies by providing training on natural gas firefighting and information on gas service to a particular address.

Today, establishing a fire brigade is a much more difficult than it was in 1911. The U.S. Department of Labor, Occupational Safety and Health Administration and the National Fire Protection Association have standards for such brigades including mandatory training and equipment requirements. The skill and danger of firefighting is no longer a casual affair that can be undertaken informally by well-meaning business owners. The role of a Corporate Fire Safety Director is of paramount importance in ensuring the safety of employees through proactive fire prevention and preparedness programs.

This Luna Park Fire Department badge may be from the amusement park's fire brigade or from one of its fire fighting attractions.

Amusement Parks

One very interesting and rare badge is from Luna Park. Luna Park, which opened on May 16, 1903, was one of many individual amusement parks in the area of Brooklyn known as Coney Island. In 1904, the park introduced a show entitled, *Fire and Flames*, in which women and children were rescued from a simulated four-story building on fire. There were literally hundreds of firemen in the "Luna Park Fire Department" who responded to extinguish the mock blaze. Luna Park opened this show just days before its rival, Dreamland, unveiled a similar one.

The fire that destroyed Dreamland Park in 1911.

A second fire attraction was added to Luna Park in 1912, depicting a western town ablaze. It was entitled, *The Great Fire Show*. Ironically, Luna Park was totally destroyed by fire in 1944.

Another Coney Island park was Dreamland. It opened in 1904 and was destroyed by fire in 1911. As mentioned, it also had a fire show entitled, *Fighting the Flames*. The show included a cast of 2,000, including 120 firefighters outfitted with four engines and hose wagons and one ladder truck commanded by Chief Sweeney. The show opened with a single six-story building and an entire city block was later added in 1905. The auditorium frequently held 1,500 spectators and the building was adorned with sculptures of firefighters. This show was replaced with one simulating the 1906 San Francisco fire and earthquake and only had a cast of 350. It is not known if any badges were ever used in connection with either of these attractions. In 1908, real New York firemen put on a show at Dreamland on Christmas Eve in which they battled a simulated fire in a burning skyscraper.

On the West Promenade of Dreamland stood "Lilliputia," better known as "Midget City" built in half-scale to resemble the city of Nürenburg, Germany. This attraction was socially acceptable at the time. Included in the 85 foot by 175 foot area was the "Midget City Fire Department." The "firemen" donned uniforms and responded to a simulated fire once an hour with a miniature hose cart and a working steamer pulled by ponies. These actors were pressed into real service using their own equipment during the 1911 fire that destroyed the park. Badges from this group were not found, but they do appear to be present in photographs of the troupe (see Chapter 23).

In June 1960, "Freedomland USA" opened in the Bronx. The 205-acre park was shaped like the United States. Each area of the map had an attraction representative of the region. In the Midwest was an attraction called, *The Chicago Fire* which was a *very* small scale re-creation of the 1871 catastrophe. A natural gas-fueled fire was successfully extinguished every hour on the hour by a troupe of uniformed actors using a real pumper. Volunteers were taken from the audience to assist and were awarded "Honorary Fire Chief" certificates. No souvenir badges from this attraction are believed to have existed.

Freedomland also had its own legitimate fire brigade. It was directed by Sidney Wohlfeld, a retired FDNY Battalion Chief and his deputy, William A. J. Beck a retired FDNY Acting Captain. When the park first opened, they operated with a 1946 Seagrave engine, acquired from the Smithtown (Long Island, N.Y.) Fire Department. Badges were not given to the brigade's seasonal members in its first year of operation. Whether or not this changed in later years is not known. A photograph of Chief Wohlfeld taken in 1962 shows him wearing a badge on his uniform, but the details of the badge are not discernable (See Chapter 23). The park failed financially and was closed in 1964.

Federal Fire Departments and Brigades

Several fire brigades have operated within the City of New York under the auspices of federal agencies. Historically, most of these have been situated at military bases and stations during World Wars I and II.

The New York Fire Department indirectly played a role in the development of the first federal firefighting units dating to the Civil War. At that time, Secretary of War, Edwin Stanton asked New York City Mayor George Opdyke to provide an experienced NYFD Engineer to become the Chief Engineer of the Fire Department of the Southern Division of the Federal Army. Engineer John Baluch was summoned and served through the war. At the end of the conflict, many Army installations maintained a fire brigade to protect the community as well as the fort. These were the predecessors of today's federal fire fighters.

Engineer John Baluch, formerly of the NYFD, as Chief of the US Army Fire Service.

Fire units have, and in some cases still do, operate at federal facilities in New York City. Badges for some of these fire units are shown here. In several cases, we know that fire units existed and they are described here even though no badges could be found. Still yet, some federal

Badges from the fire department at the U.S. Coast Guard base on Governor's Island, NYC.

agencies are believed to have had fire units but since no documentation could be found, they are not included in this chapter (e.g., the famous U.S. Immigration facility at Ellis Island). Contemporary federal agencies with some form of fire safety or fighting units include the National Park Service, the National Passenger Railroad Corporation and the U.S. Coast Guard Fire Department on Governor's Island.

U.S. Coast Guard Stations

Governor's Island is situated one-half mile southeast of Manhattan. It was purchased from the Canarsie Indians in 1698 and served as a home for the colonial Governor. After the Revolution, the U.S. Army occupied the site until 1966 when it was turned over to the U.S. Coast Guard to serve multiple commands for the metropolitan area and the northeast seaboard. The U.S. Coast Guard station at Governor's Island has the distinction of being the oldest continuously operated military base in the United States. It is federal, not State or City property.

A fire department comprised of fifteen firefighters manning one truck and two engine companies protects the island. All personnel are civilians. The Coast Guard also operates a crash-fire-rescue vehicle at its air base at Floyd Bennett Field in Brooklyn. The FDNY does not respond to fires on Governor's Island unless called by the Coast Guard Fire Department dispatcher. This is done by activating FDNY alarm box 8204, in contrast to the pre-war period when "a blank charge will be fired from the 'reveille gun' and in addition, the Manhattan Central Office will be notified by telephone" to summon the fire department. Vehicles responding to the island would have to be transported by ferry from the Coast Guard's dock in Manhattan.

Fire Department at the Manhattan Beach (Brooklyn) Coast Guard Station during WWII.

Staff of the Governor's Island Fire Department

In 1997, the Coast Guard moved its operations off the island to a variety of other locations. The fire department remains to protect the 225 buildings spread over the 178-acre island. Six buildings have been granted historic landmark status, including the Governor's house built in 1708. Although the destiny of the facility is yet to be determined, the fire department expects to remain on-site at least until 2002. During this period, mutual aid is provided under a formal Memorandum of Understanding between the Coast Guard and the FDNY which specifically details the type and method of supplemental firefighting support that the Department will provide.

The New York Port of Embarkation Fire Dept was operated by the U.S. Army during World War II. (Photo Courtesy of Wayne Sorenson)

Over the years, several badge styles have been issued to USCG fire departments including Governor's Island. Several are shown here. It is important to note that the only ranks that exist at Governor's Island are Fireman, Captain, and Chief.

The Coast Guard also operated a fire department at its Manhattan Beach installation in Brooklyn. Although no records other than photographs could be found, apparently the brigade was staffed by Coast Guardsmen who were not issued any fire fighting badges or identification other than turnout gear. In 1942, U.S. Maritime Service opened a Training Station on the seventy-six acres adjacent to the Coast Guard Station and established their own fire department, as discussed below.

U.S. Army Installations

On July 31, 1916 saboteurs led by Franz von Papen set fire to the Black Tom pier in Jersey City, New Jersey where some two million tons of black powder were stored in thirty-two freight cars and ten barges awaiting shipment to troops fighting the First World War. The fire raged for two weeks and six persons were killed. When munitions began exploding, the Statue of Liberty located one-quarter mile away in New York harbor was bombarded. One shell landed directly down the stack of the fireboat *Strong*. Fortunately, it was a dud. Fire protection for the Port of New York was provided by the Army's 301st Guard and Fire Company. A Lieutenant in that unit, John J. McCarthy, an FDNY fireman on military leave, was appointed Fire Marshal for the New York Port of Embarkation. Lt. McCarthy later went on to become Assistant Chief of Department in Command in the FDNY and was the author of several books on wartime firefighting. In 1917, a fleet of tugboats was organized as a fleet of auxiliary fireboats. These fleets hailed from the railroads of New York and New Jersey, Bush Terminal and the Socony Company. Throughout the war, one FDNY fireman was stationed on every barge in the harbor or at dock that contained ammunition or explosives. While there were numerous pier and ship fires during the war, none were as devastating as the New Jersey fire.

Several fires did occur at New York piers during the Second World War, although none were attributed to sabotage. A five alarm blaze ripped through Pier 27 in Brooklyn on August 18, 1941, killing thirty-four, injuring seventy, and requiring the largest response to a single fire in the Department's history up to that time. On February 9, 1942, the *Normandie* caught fire while being converted from a luxury liner to a troop carrier, and rolled on its side at Pier 89 in Manhattan.

To increase fire protection at the piers during this period and to supplement the depleted force of the FDNY, the U.S. Army's New York Port of Embarkation (NYPE) established fire departments at its principal terminals. The NYPE operated a total of ten terminals and three areas for staging troops throughout NYC staffed by approximately 55,000 military employees. The men who staffed the fire department were often not qualified for military duty but wanted to do their part for the war effort as civilian employees in the U.S. Army.

The New York Port of Embarkation Fire Department badges.

The badge issued by the NYPE was a standard Maltese cross with the letters, "NYPE" in the upper arm, "F.D." in the lower, and applied numbers in the center.

The U.S. Army has had installations in New York City since colonial times. Although records are not available in all cases, it is believed that all had some form of fire protection at least during the Second World War. These installations include Fort Totten in Queens, Forts Tilden and Hamilton in Brooklyn and Fort Wadsworth in Staten Island. Consistent with federal firefighters at the time, these units were manned by uniformed military personnel, with the exception of the New York Port of Embarkation. It is probably for this reason that no badges could be directly connected with any of the U.S. Army forts in New York City.

Bush Terminal in Brooklyn had fire protection provided by the U.S. Army during its occupation of the site and by a private brigade in the years after.

U.S. Maritime Service Training Station

The U. S. Maritime Service Training Station was a seventy-six acre installation situated in Manhattan Beach, bounded on three sides by water. The Station fire department was commanded by Chief Boatswain's Mate John G. Buckley, U.S.N.R., who was an FDNY dispatcher prior to the war. A fire house at the Station housed a 1942 Mack pumper and five trailer pumps. The department was comprised of twenty-five men from the Navy and Merchant Marine including former members of the FDNY and other departments around the country. All received military fire training from the U.S. Navy Firefighters School. There were twenty-three alarm boxes strategically located around the Station and a Class-3 building box at the Housewatch Desk in the fire house.

The badge from the Maritime Station is a simple stock badge with eagle bearing the letters, "U.S.M.S." and the words, "Fire Dept."

Rare badge from the U.S. Maritime Station on Manhattan Beach, Brooklyn during World War II.

Merchant Marine firefighters subdue a blaze in the hold of a vessel as part of their training at the U.S. Maritime Service base at Manhattan Beach, Brooklyn during WWII.

U.S. Naval Air Station - New York

A small island off the south shore of Brooklyn near Flatlands, known as Barren Island, was used as a dumping ground by the City. It was inhabited and had its own fire department whose badge is shown in Chapter 3. In 1930, the City's first municipal airport, Floyd Bennett Field, was opened and a bridge connected it to the mainland. In 1939 all inhabitants were ordered off the island by the City for the development of the Belt Parkway. In

The Naval Air Station - New York was originally the site of New York City's first municipal airport, Floyd Bennett Field on Barren Island.

1942, the Navy purchased the island and took over the airport as the Naval Air Station - New York. The Navy operated this site until 1971 when it was given back to the City which used it for a variety of purposes including as an air station for the NYPD Aviation (helicopter) Bureau. The U.S. Coast Guard occupies a segment of the site and, as noted earlier in this chapter, operates a crash-fire-rescue unit there.

Only two badges from the Naval Air Station - New York fire department, a breast and cap badge without rank, were located. The breast badge is a custom round badge circled by a hose and topped with an helmet. The cap badge is a small Maltese cross with the traditional fire scramble in the center. It is not known if any other styles or ranks were ever issued.

New York Naval Shipyard

A fire department established at the New York Naval Shipyard, better known as the Brooklyn Navy Yard, operated for about 100 years. The Navy first acquired the property on the Brooklyn waterfront in 1800. Many historic vessels were outfitted there including the *Monitor* during the Civil War, the infamous *Maine* of the Spanish-American War, and the two ships that respectively symbolized the beginning and end of the Second World War, the *Arizona* and the *Missouri*. More than 71,000 people worked at the Yard during World War II. Many years later in December 1960, fifty people were killed when a fire destroyed the aircraft carrier, *Constellation* at the docks of the Navy Yard. The City purchased the majority of the property in 1965 and converted it to an industrial park, with fire protection currently being provided by the FDNY.

Very early badge from the Brooklyn Navy Yard, probably mid to late 1800's.

Stereoview of the Brooklyn Navy Yard Fire Department circa 1870.

The Brooklyn Navy Yard, officially the New York Naval Shipyard, had its own fire protection from it's earliest days until after the Second World War.

It is not certain when the first fire brigade, or department, was formed at the Navy Yard. A newspaper account of the department testing a new chemical engine appeared in *Harper's Weekly* in January 1874. A badge located for the Navy Yard Fire Department is consistent with the style issued to the Brooklyn Fire Department - Western District during the 1855 to 1869 period. The Navy Yard was located in the Western District. Although the Navy Yard badges could have been introduced *after* the Brooklyn Fire Department stopped using this design, it still suggests that the Navy Yard had a fire brigade at least as early as the mid-1800's.

Badges issued to members of the New York Naval Shipyard Fire Department during the World War II era consisted of a Maltese cross with the letters, "N.Y.N.S." in the lower arm and the rank or title in the upper arm. Several versions of badges for Fireman can be found. The difference is in the way the central disk is treated. It is not unusual for there to be red or black paint added in the central disk. This is clearly a modification made after manufacture since the paint is not a hard-fired enamel. It appears that members of the N.Y.N.S. fire department added these colors, possibly to distinguish between engine and ladder company members. As is well known, the FDNY has used these distinguishing colors on helmet frontpieces for over 150 years; black for engine companies, red for ladder companies. Badges found without this coloring might have been stripped off, since many collectors thought the coloring was added by someone outside of the department.

National Park Service

The National Park Service of the U.S. Department of the Interior manages several sites within the City such as Federal Hall National Monument, Theodore Roosevelt Birthplace National Historic Site, General Grant National Monument, Castle Clinton National Monument, Statue of Liberty and Ellis Island National Monument and the Gateway National Recreation Area. Gateway National Recreation Area encompasses over 26,000 acres of property spread along the Brooklyn and Queens south shoreline as well as sections of Staten Island and Sandy Hook in neighboring New Jersey. At the Brooklyn segment of the site, Park Rangers staff two brush fire trucks. Park Rangers undertake typical duties and are on-call for fires. They do not wear any distinctive fire badge (only their NPS Ranger badge) nor any distinctive emblems.

City-Agency Fire and Related Services

NYC Department of Buildings

After the March 1911 Triangle Shirtwaist Company fire and the subsequent investigations into factory conditions, the Sullivan-Hoey law was passed in October of that same year. The law stepped-up fire prevention activities in the City by increasing the authority of the Fire Commissioner and establishing the Bureau of Fire Prevention within the Department. In 1933 however, the duties of the Bureau were transferred to the NYC Department of Buildings. A unique style badge was issued to members of the Department of Buildings unit. The badges have a shield design with two torches on either side and the City seal in the center. An upper banner identifies the bearer's title while the lower banner says, "Bureau of Fire Prevention." Many fire collectors are confused by this badge since it does not say, "Fire Dept. New York" although the Bureau is, and was, a part of the FDNY. The Bureau returned to the FDNY in 1940 and is still an active part of the Department with both uniformed and non-uniformed personnel.

The Triangle Shirtwaist Company Fire, March 1911, caused the enactment of the Smith-Hoey Law which mandated the establishment of the Bureau of Fire Prevention.

The NYC Department of Buildings managed the Bureau of Fire Prevention from 1933-1940.

NYC Department of Corrections

Numerous jails are operated by the NYC Department of Corrections (DOC), including the maximum security prison at Riker's Island. Riker's Island consists of 415 acres in the East River, just west of LaGuardia Airport. It was first established as a prison farm in 1884 and now houses 15,000 inmates. A fire brigade is operated on the Island by the NYC DOC, manned by Correctional Officers. A smaller unit equipped with a former 1934 FDNY 1,000 gallons-per- minute Ahrens-Fox pumper, was once stationed on Hart Island in the Long Island Sound. It served as a reformatory prison but is also the site of the City's infamous, "Potter's Field" with 750,000 interments since 1869. Due to obvious security restrictions, badges and insignia worn by Correctional Officers could not be photographed for publication.

1981 Mack pumper (former FDNY Engine 277) operated by the NYC Department of Correction on Riker's Island. (Photo courtesy of Michael Martinelli.)

NYC Department of Sanitation

Garbage is a major challenge for New York City. More than 26,000 tons of garbage are generated each day and the battle over how to dispose of it is never-ending. The City's last landfill, Fresh Kills landfill on Staten Island, is the world's largest landfill, and it is the highest point of land in the City. Although it was shut down in March 2001, debris from the World Trade Center collapse was sent there from September 2001 through May 2002. For many years, the Department of Sanitation maintained firefighting apparatus at this and other landfills around the City. The apparatus was used to extinguish small fires that are inevitable in this type of facility, or to mount an initial attack while awaiting the response of the FDNY. Workers from the landfill would staff the equipment in case of emergency.

Apparatus operated by the NYC Department of Sanitation at one of three landfill facilities that had fire brigades until the 1970's. (Photo courtesy of John A. Calderone.)

State Fire Services

Manhattan State Hospital

On isolated Ward's Island, situated in the Harlem River, is Manhattan State Hospital. The hospital was established in 1899 upon the consolidation of two other hospitals on the island. At the time, it was the largest psychiatric center in the world with 4,400 beds. Until construction of the Hells Gate bridge in 1917, access to the island was limited to ferry boats. In 1923, a major fire at the facility killed twenty-two patients and was the impetus for major safety reforms in psychiatric hospitals throughout the state. A fire brigade was formed and manned by public safety officers. In 1925, the brigade was equipped with an Ahrens-Fox tractor-drawn aerial ladder, which was replaced in the 1940's with a Mack aerial ladder. The brigade at Manhattan State Hospital no longer exists. The hospital now consists of many multi-storied buildings providing more outpatient services than in-patient care; there are presently 850 beds. Access to the island is now possible from Queens, Manhattan, and the Bronx by way of the Triborough Bridge, which was constructed in 1936.

The Manhattan State Hospital fire brigade consisted of officers of the NYS Department of Mental Hygiene Safety Services. Their badge was a sun-ray Maltese cross with eagle, with a reverse blue enamel ring displaying the agency's name and the words, "Safety Services." The upper tab displays the rank, a lower tab with a badge number, and the State seal is in the center.

It is believed that the Willowbrook State School on Staten Island, the site of a treacherous fire in its subterranean tunnel system in 1952, also had similar protective fire services. But, this could not be confirmed. If it existed, the same badge probably would have been used. This school for mentally ill children opened in 1952 and sparked much controversy in 1971 when overcrowded and inhumane conditions were revealed by the media. It was finally shut down in 1982.

The New York State Department of Mental Hygiene operated fire brigades at the Manhattan State Hospital on Ward's Island and Willowbrook State Hospital on Staten Island.

Transportation-related Fire Services

Entry to and exit from New York City can be made via its three international airports, over bridges, through tunnels, at bus and railway terminals, and even by ship. Many of these facilities and structures are owned and/or maintained by the Port Authority of New York and New Jersey, the Metropolitan Transportation Authority and its subsidiaries, the Triborough Bridge and Tunnel Authority, the Long Island Railroad, Metro-North Commuter Railroad, and the National Passenger Railroad Corporation. Although all are protected by the FDNY, many of these facilities maintained their own fire brigades in the past and some continue to do so.

Triborough Bridge and Tunnel Authority

In 1939, the FDNY shops converted five pumpers into "Bridge Chemical Units" specially equipped to fight vehicle fires on the bridges connecting various boroughs of the City. These units were retired from service in 1959. Emergency road services at the over-water and underwater vehicular accesses within the City is provided by the Triborough Bridge and Tunnel Authority (TBTA). This agency, known previously as the Department of Bridges is now a part of the Metropolitan Transit Authority (MTA), a public authority established by the State of New York in January 1968. Fire apparatus is not operated by the agency. Rather, emergency road service vehicles are equipped with some fire equipment such as extinguishers and booster lines for fighting incipient stage fires. The FDNY is called for response to all fires at TBTA crossings. TBTA emergency road service personnel are not issued badges.

Port Authority of New York and New Jersey

The Port Authority of New York and New Jersey is a unique bi-state public agency. At the airports, the Port Authority Police provide primary fire response with simultaneous notification to the FDNY. (It should be noted that the Port Authority also covers facilities outside of New York City, where it interacts similarly with other local fire departments.) This is an evolution from the early days of fire protection for the airports when, in 1939, a "Crash Wagon" was put in service at LaGuardia Airport. The single piece of apparatus was owned by the Dock Department, predecessor to the Port Authority, and was manned by two FDNY members special-detailed from rescue companies. During the war years, the U.S. Army Air Force provided fire protection at LaGuardia. In 1946, it turned this responsibility and several pieces of apparatus over to the FDNY which organized the Airport Crash Unit. The Department relinquished this role to the Port Authority soon thereafter in 1947.

Bridge Chemical Unit built by the FDNY in 1939 and discontinued in 1959. (Photo courtesy of John A. Calderone.)

FDNY Crash Wagon at the NY Municipal Airport (later LaGuardia Airport) in 1939. (Photo courtesy of John A. Calderone.)

The Port Authority of NY and NJ provides primary fire response at the two international airports in New York City; JFK and LaGuardia.

Aircraft firefighting foam truck operated by the Port Authority of New York and New Jersey Police at LaGuardia Airport in 1999. (Photo courtesy of Lawrence Rand.)

John F. Kennedy International Airport, which opened in 1948 as Idlewild International Airport, covers nearly 5,000 acres in southern Queens and is virtually a city unto itself. Just as it does at LaGuardia Airport, the Port Authority operates crash-fire-rescue apparatus. The Port Authority Police are first due at all fires and emergencies at both airports.

All officers of the Port Authority Police receive initial training in police skills and firefighting. However, at the three New York metropolitan area airports, where firefighting vehicles are staffed, officers are permanently assigned to this duty and receive additional, specialized training. Since all such firefighters are uniformed Police Officers, they wear a regulation Port Authority Police badge. However, a distinctive embroidered patch is issued.

Mayor Rudolph Giuliani proposed taking control of the two airports away from the Port Authority and placing them under a New York City Airport Authority. If such an agency were to be created and the City takes over operation of the airports, police and fire service will be provided by the NYPD and FDNY respectively. This would be the first time that the FDNY would operate airport crash-fire-rescue units since the original Airport Crash Wagon of the 1940's. As of this writing, this proposal has been tabled.

The Port Authority also covers connections between the City and New Jersey, including the Lincoln and Holland Tunnels, the George Washington, Bayonne, Goethals Bridges and the Outerbridge Crossing. The state dividing line is situated at the mid-point of each crossing but the Port Authority has jurisdiction in both as an unique bi-state agency. The Port Authority operates emergency service vehicles similar to those of the Triborough Bridge and Tunnel Authority, equipped with fire extinguishers, booster lines and firefighting agents such as Purple K and Halon. The operators of these vehicles are called Bridge & Tunnel Agents. The Agents are fully trained in all aspects of firefighting, extrication, first aid and hazardous materials response. They are issued uniforms with a badge and a distinctive embroidered emblem. Two badges are issued, one for Tunnel & Bridge Agents and a second for Senior Tunnel & Bridge Agents, or "Crew Chiefs."

Early emergency truck, equipped for fighting fires in the Holland Tunnel connecting New York City and Hoboken. (Note the fire bell on the truck and its ability to be driven from either end.)

The Port Authority of NY and NJ provides fire protection at the Hudson River crossings between the two states.

Current emergency truck operated by the Port Authority. (Photo courtesy of Michael Martinelli.)

Railroads

Back in the days of steam locomotives, many railroads across the country maintained their own fire departments. Fires were an ever-present danger on the trains themselves, along the tracks, and in work yards. Many of these fire brigades were equipped with specially designed rail cars for firefighting and for transporting firefighters along the tracks. Prior to consolidation, New York City and its surrounding communities were served by many railroads. Most of these evolved into one or more of the current rail lines that transport hundreds of thousands of residents, commuters, and tourists in and around the City every day.

Badges from several early railroad fire departments that had operations in New York City are shown here. The Pennsylvania Railroad served NYC since the early 19th century. It acquired a majority of the stock in the Long Island Railroad in 1900. The Lehigh Valley Railroad transported much needed and highly combustible coal from western Pennsylvania to New York City. The New York Central and Hudson River Railroads (NYCHRR) were consolidated in 1869 by Cornelius Vanderbilt, whose statue stands guard over the entrance to today's Grand Central Terminal. Some lines continued to operate under their original names while others took on the consolidated name. In 1968, the NYCHRR merged with the Pennsylvania Railroad to form Penn Central RR which subsequently became a part of the National Railway Passenger Corporation (Amtrak) in 1971.

Many railroads that have served New York City operated their own fire departments.

The present intra-City rail system is its famous, or infamous, subway. Fire service for the subway is provided by the FDNY. Fire badges with the initials, "NYCRR" on them are *not* from the New York City Railroad, but the *New York Central* Railroad, as previously described.

Four major railroads continue to service the City; three as commuter lines from the outlying suburbs plus the national rail system. While there are hundreds of entry and exit points into the train and subway systems, two stand out. They are Grand Central Terminal and Pennsylvania (Penn) Station.

Grand Central Terminal is a New York City landmark, first built in 1869 as Grand Central Depot for the New York Central and Hudson River Railroad. The current building was completed in 1913. The main concourse is a huge space, 120 feet wide, 375 feet long, and 125 feet high. There are over 200 railroad and subway tracks within its structure. A multiple-alarm fire occurred on August 27, 1985 destroying eighteen train cars in a tunnel on the third underground level, 100 feet from the station. This incident prompted the formation of the current fire brigade at the Terminal and the hiring of a full-time Fire Safety Director. Because of their expertise and knowledge of the terminal and the operation of the trains and other equipment, the Metro-North Fire Brigade at Grand Central Terminal is an important resource for the FDNY upon response to any fire or emergency.

Given its landmark status and labyrinth of subterranean tracks, it is not uncommon for the Terminal to be used as the location for movie and television scenes. When such filming takes place, the fire brigade stands by. During the 1996 filming of the movie *Extreme Measures*, several scenes were shot along the labyrinth of abandoned tracks with the star carrying an open-flame torch for lighting. A accidental fire occurred during the filming of a gun battle scene for the movie *Carlitto's Way* in 1994, when barrel flashes from the blank rounds of ammunition ignited grease in an elevator pit.

Metro-North Commuter Railroad is the second largest commuter rail line in the country, operating what was formerly the New York Central Railroad and the New York, New Haven and Hartford Railroad. The system is operated by the MTA. While covering multiple fire jurisdictions, almost all of the lines converge at Grand Central Terminal in New York City. Metro-North maintains fire brigades at Grand Central Terminal as well as several of its facilities outside of New York City at Croton-Harmon and Brewster. The brigades, like that of Amtrak at Penn Station, consist of workers in the terminal and

Aftermath of a fire at the Bowery station of the New York City subway system fought by the FDNY.

Members and equipment of the Metro-North Fire Brigade at Grand Central Terminal.

Penn Station also serves as the Manhattan terminal for the Long Island Rail Road (LIRR). LIRR is the third oldest railroad and the largest commuter rail line in the country. It is the only one still using its original name. The line provides commuter service to Penn Station from Long Island, which lies east of the City. There are also terminals at Flatbush Avenue in Brooklyn and Hunter's Point Avenue in the Bronx. The current LIRR is the result of a long history that began with its first train run on April 18, 1836 and has included mergers with numerous other local rail lines. Diesel or electric trains run on 700 miles of track. The system is operated by the MTA.

LIRR has a staff of fire marshals as part of its Safety Department. The fire marshals provide fire safety training to LIRR employees, perform routine fire inspections and operate as industrial fire brigades as defined by OSHA. The brigade at Grand Central is comprised of fifty to seventy-five fully trained Metro-North employees who volunteer for brigade duty. They are equipped with a full complement of firefighting gear mounted on three electric carts. The fire brigade also responds to medical emergencies within the Terminal with a specially-equipped emergency medical cart. Until the 1940's, such service was provided by two doctors who staffed the Grand Central Terminal Emergency Hospital.

Badge designs for the Metro-North Fire Brigade are at the discretion of the Chief. Therefore, numerous designs have been used over the years, most of which are based on stock badges from manufacturers. These include a standard Maltese cross design, sunburst police style, and other generic shapes. At the present time, badges for Chief and Assistant Chief include the officer's name. Although participation in the fire brigade is voluntary and secondary to the individual's principal job with the railroad, uniforms are worn when the brigade participates in formal ceremonies, parades, etc. Therefore, hat badges and breast badges are issued. The cap device for brigade members is a Maltese cross. Cap devices for officers are similar to the FDNY trumpet pattern.

Metro-North Railroad protects the Grand Central Terminal, a NYC landmark, as well as its tracks running into the City.

investigate any fires that occur. They work closely with the numerous fire departments that provide fire protection along the three track lines that run the entire length of Long Island.

In addition to the LIRR, the National Railroad Passenger Corporation, better known as "Amtrak," provides services to New York via Pennsylvania (Penn) Station. It operates train service throughout the country including a popular service known as the "*Metroliner*" that takes passengers from Penn Station in NYC to Union Station in Washington, D.C. within three hours. The line, formed in 1971, is a federal agency.

The National Passenger Railroad, or AMTRAK, operates 76 miles of track within New York City including its terminal at Pennsylvania Station.

Long Island Railroad protects its 700 miles of tracks as well as its stations and yards.

The first Pennsylvania (Penn) Station was built by the Pennsylvania Railroad in an attempt to compete architecturally for rail business with the successful Grand Central Terminal. Construction began in 1902 and the terminal opened in 1911. In 1962, the structure was torn down and the outrage it caused was the impetus for the important, and still very active, City's Landmark Preservation Commission. The new station was located in the lower levels of the office complex that also houses Madison Square Garden. It is anything but ornate. When the U.S. Postal Service announced that it no longer needed its main New York City post office located across the street from and above Penn Station, Amtrak acquired the space and plans to remodel it for a new, more elegant station. The Post Office building was designed at the same time and by the same architects of the original Penn Station. Today, more than half-a-million passengers utilize Penn Station on their journeys into, out of, and through New York City.

Fire apparatus of the AMTRAK Metropolitan Division fire brigade in 1999. (Photo courtesy of Michael Martinelli.)

Amtrak has a fire brigade at Penn Station which consists of Amtrak employees in numerous jobs who respond to fires as needed. They are not equipped to fight anything beyond incipient fires using portable 125-pound ABC extinguisher systems. These Amtrak employees are identified only by a decal on their hard-hat; no badges are issued. At the large Sunnyside yards in Queens, a mini-pumper on a 1998 GMC chassis is equipped to run on the rail tracks. The Manager of Fire Safety, who is responsible for fire prevention along more than seventy-six miles of track, does have a badge as shown.

The Port Authority Trans-Hudson Corporation (PATH) was formed in 1962 utilizing fourteen miles of track formerly operated by the Hudson and Manhattan Railroads. These tracks were originally laid between 1874 and 1909. Trains from New Jersey terminate at Penn Station and formerly terminated at the World Trade Center prior to September 11, 2001. As its name implies, PATH is operated by the Port Authority of New York and New Jersey which, as previously described, has its own police department. PATH does not operate fire brigades at its New York terminals. The World Trade Center station was a small facility protected by the FDNY and Penn Station has a fire brigade operated by Amtrak.

CHAPTER 22

Unidentified Badges

While searching for badges for this book, many were found that are thought to originate in New York City but for which no documentation or verification of their use could be found. These badges are in the hands of collectors and museums.

Of these unverified badges, one of the oldest is believed to be a copy of an 1860 engine company badge. The basic badge is quite similar to the ones actually issued. The pin on the reverse side suggests that this is a period piece. The most confounding feature of the badge is the use of a roman numeral for the company number instead of an arabic number. This suggests that it is not a Department-issued badge. The FDNY has never used roman numerals on any of its badges. However, Engine 12 *did* use roman numerals on their helmet frontpieces and on their engine. During this era, the badge and badge number would have been in opposite color metals, namely Prince's metal and white metal. On this badge the numbers and the badge are both made of Prince's metal. This badge may have been issued to an honorary member of the company, or perhaps was a retirement piece or the first "saver" badge.

1860-period badge with roman numerals in lieu of typical arabic numerals for company number. This may be a badge for an honorary member.

One extremely interesting badge is believed to be from around 1870. It is an ornate cross with the intertwined letters, "NYFD" in the center that were typical of other badges made in the 1870's. Four of the five examples of this badge that were examined are numbered, some in the 200 range. The fifth appears to have had the number removed. All are made of brass. These features suggest that the badge may be an early Chief officer's badge. However, early Department Orders that specify badges do not describe this badge even though they do describe Chief officers' cap devices. The NYFD designation was not used during the 1860-1865 Metropolitan Fire Department period, and the intertwined letters are not seen on earlier badges. This supports the conjecture that the badge is circa 1870.

Badge believed to be that of a Chief Officer in the 1870-1884 period.

A badge in the same style of the Engineer of Steamer badge with the word, "Driver" in place of Engineer was found. Unlike the Engineer of Steamer badge, it has the letters "FDNY" engraved. Most FDNY historians and badge collectors believe this to be a fake badge. The title Driver was not used after the horse-drawn era and this badge displays a motorized vehicle of a 1920's to 1940's vintage. Instead, it would have been used during a time when the title, Chauffeur was used and for which a pentagon-shaped badge was issued. Chauffeurs were ununiformed civilian personnel, whereas this badge has a screw and post for cap mounting.

"Driver" badge with FDNY designation. May have been made personally by a civilian employee to be worn on a uniform cap while driving Department officials.

In 1899, Commissioner Scanlon authorized badges for retired firemen. It was common at that time for retired members to be referred to as "Veterans." A small badge with the word, "Veteran" in the center ring above the number "1" was found. Given the number and the absence of any other such badges, it is believed that this badge may have been a prototype for the badge issued to retired members. As described in Chapter 13, the badge that was issued to retired members is believed to be the standard pentagon-shaped badge with the individual's name engraved in the upper banner.

It is believed that this is a prototype badge.

Badge thought by some collectors to be from the Brooklyn Fire Department, but no documentation could be found to verify this claim.

A badge displaying the letters, "B.F.D." and the word, "Exempt" is believed by many collectors to be from the Brooklyn Fire Department. The word Exempt suggests that it would date to the period of the volunteer fire department. However, no other badge in this particular frontpiece style is attributed to the Brooklyn Fire Department. This frontpiece style was used by many other fire departments in the New York City area.

A Maltese cross bears an NYFD designation in the lower arm displays the letters "V.S." in the upper arm and the letters "H" and "L" engraved on either side of an applied number "2" in the center disk. It is a cap badge. It may be a badge made by or for a member of the WWI-era Volunteer Emergency Service or Auxiliary Corps. However, the letters "V.S." are cause for uncertainty. One would expect the letters, "V.E.S" or "A.C,." both of which were used in different applications during the 1918-1919 period. It has been suggested that this may be an early badge from Valley Stream, NY Fire Department. Valley Stream is on the eastern border of New York City.

Badge with NYFD designation but not consistent with other Volunteer Service badges.

The title, "Tour Chief" was not likely in use during the 1884 to 1937 period when the pentagon badge bearing this title was common. Given its silver color, if "Tour Chief" is a legitimate title then it would not be used for a Chief Officer, but rather for a civilian post. Again, it is an unlikely title for a civilian employee of the period. In the absence of any confirmation of this title, the badge is presumed to be illegitimate.

The title "Tour Chief" is not known to have been used during the 1884-1937 period.

Many company names were common to hundreds of fire departments other than the FDNY and there are many badges that collectors might claim to be of NYC origin but for which there is no documentation. Examples include Pioneer Hose Company 1, Union Company 1, Ever Ready Engine Company, Washington Company 1 and Neptune Engine Company badges.

There was a Pioneer Hose Company within the Newtown Fire Department, but with such a common name as Pioneer, there is no certainty that this badge is from there. Similarly, there were four Union Company 1's among the early volunteer fire departments in New York including East New York, New Lots, Gravesend and College Point. The base badge is similar to that used by companies throughout the area in the nineteenth century which leads to the assumption that this badge originates from one of these departments.

These two badges are in the style of Long Island City FD badges but display company names that are not consistent with companies from that department.

Badges with company names that may be from NYC.

Only one company in New York City used the name Ever Ready. It was a ladder company of the Springfield Fire Department. But, the badge clearly shows a steamer. The base badge is the same as the one used by the Long Island City Volunteer Fire Department, but no company by that name has been documented in Long Island City.

Washington was a common name taken by fire companies. Those with a company designation of "1" in New York have included Engine 1 of the Brooklyn Western District, Flatbush, North Shore and Eastchester Fire Departments, and Ladder 1 of the Flatbush, Charleston/Kreischerville, and Edgewater (Staten Island) Fire Departments. The badge shown with "Washington 1" could be from any of these. A second badge with the name Washington is obviously from a ladder company in a shape associated with the Long Island City Fire Department. But there is no record of a Washington Ladder Company 1 in the volunteer or paid Long Island City Fire Department.

Two badges that are from a Neptune Engine company are shown. In both cases the number one seems to suggest the company number. There was an Engine Company 1 in the New Lots Fire Department. The style of one of these badges is like the badges used by the Long Island City Fire Department. But the Long Island City Fire Department did not have a company named Neptune. The number might be a badge number rather than a company number. If the number "1" is not a company number, then the badges may be from one of the other engine companies in the New York area that had the name Neptune. The other Neptune Engine Companies included Engine 6 in New York City and Edgewater, Engine 2 in Brooklyn, Bay Ridge and Jamaica, Engine 7 in Brooklyn Eastern District, and Engine 3 in Riverdale.

Anyone with information or documentation on these or any other badge believed to be from New York City are encouraged to contact the author through the New York City Fire Museum.

CHAPTER 23

In Search Of...

Although many readers are astounded by the number of different badges that have been located and photographed for this book, there are still many that are known or believed to exist but were not found. These might be in the archives of museums, in the hands of private collectors or possibly held by descendants of their original owners.

Of the badges known to have existed, there are several one-of-a-kind, custom pieces that may or may not have survived the past century; their owners may have taken them to their graves. The custom badges of two Commissioners, Alexander Shaler of New York and John Ennis of Brooklyn are illustrated in the 1887 books, *Our Firemen* and *Our Firemen (Brooklyn)* respectively, but were not located. Another one-of-a-kind piece that may still exist is the first New York City Fire Marshal's badge. This badge and subsequent Fire Marshal badges were issued by the Police Department until the position was brought into the Fire Department in 1873. Fire Marshal badges from the 1854 to 1873 Police Department period and from the 1873 to 1884 Fire Department period have yet to be found.

Honorary Medical Officer and Deputy Commissioner Harry M. Archer.

Two historic badges that have not been found. Metropolitan Fire Department Commissioner Alexander Shaler (left) and Brooklyn Fire Department Commissioner John Ennis (right.)

Many badges were issued during four historic periods of the Department: prior to 1855 when badges were issued by individual companies, not the Department; 1855 to 1865 when firefighters' badges went through three distinct style groups but few higher-ranking badges have been found; 1865-1870, the Metropolitan Fire Department era; and 1870-1884 when the organization and issuance of badges really took hold. Any badges in addition to those illustrated here would add greatly to the documentation of these periods of the Department's history.

Badges owned by legendary figures such as Chief Harry Bonner and Dr. Harry M. Archer or legendary buffs such as Mayor Fiorello LaGuardia and Governor Alfred Smith would be of great interest.

One of the captivating aspects of researching and collecting NYC firefighting-related badges is the interesting

A complete collection of NYFD Volunteer Emergency Service badges from Fireman through Chief. The status of this collection and whether or not it remains intact, is not known.

abundance of badges for non-firefighting positions. Many are displayed here in Chapters 8 and 10. Badges from pre-consolidation departments, most notably Brooklyn would be of interest as would badges for the Medical Officers and Surgeons of the New York Fire Patrol.

A black-and-white photograph of a complete set of 1918-1919 NYFD Volunteer Emergency Service badges exists, but the collection has reportedly been broken up. The high-rank badges of Chief and District Chief, as well as the company-rank of Sergeant would complete the color photographs of the other badges that have been included here.

An example of the badge issued to the Municipal Explosives Commission is sought. It is reported to be a gold, pentagon-shaped badge bearing the words, "Municipal Explosives Commission" in reverse blue enamel in the top banner and "Fire Dept. New York" in the lower banner. It is only three-quarters the size of a standard FDNY pentagon badge. The Municipal Explosives Commission was formed in 1902, after a disastrous dynamite explosion and fire in the Murray Hill section of Queens. It was responsible for promulgating rules for the manufacture, storage, sale, transportation and use of explosives, combustibles and chemicals, which were enforced by the FDNY Division of Combustibles. The Commission was abolished in 1914.

This badge from the Standard Oil Company plant on the west shore of the Hudson River in Bayonne, New Jersey suggests that badges may also have been used at the SOCONY plants in New York.

As described, there have been numerous firefighting entities within New York City other than the FDNY. Information on some of these has been difficult, if not impossible, to find. Any additional information, including badges would contribute to the knowledge of New York's firefighting history. Badges from the Socony facilities and Ellis Island, among others, are believed to have existed. Two relatively recent badges thought to exist are from

The Midget City Fire Department at Dreamland amusement park is seen here in full uniform including, what appear to be, cap badges.

Thomas O'Brien, as Chief of the 1964 New York World's Fair Fire Department and a Captain's badge from this unit.

One of the more unusual firefighting-related badges that may have existed is from the "Midget City Fire Department" attraction at Coney Island's Dreamland. This department wore uniforms and manned miniature working apparatus. Photographs from the era show some type of badge on the cap of the firefighters. A photograph of Chief Sidney Wohlfeld of the Freedomland Fire Department shows him wearing a badge. It would be interesting to find Chief Wohlfeld's Freedomland badge.

Dreamland also had firefighters at its Fighting the Flames *attraction.*

Chief Wohlfeld promoting Fire Prevention Week at Freedomland in 1962.

If any reader knows the whereabouts of any of these or other FDNY-related badges or any photographs clearly showing a badge of New York City origin, please contact the author through the New York City Fire Museum so that an historic record can be made of them.

Appendix

EXCERPT FROM

AN ACT for the better regulation of the Firemen of the city of New York
Passed March 29, 1855

The People of the State of New York represented in Senate and Assembly, do enact as follows:

§ 16 The Common Council shall, in addition to the fire cap now in use, design a badge, and of which shall be worn by every fireman when on duty as such, and by such exempt firemen as the Board of Commissioners may give permission to in writing; and they shall pass such ordinances as may be requisite, to prevent the approach of persons other than firemen or policemen to the vicinity of fires.

§ 17 Any person who shall falsely represent any of the members of the Fire Department of the City of New York, or who shall maliciously, with intent to deceive, use or imitate any of the signs, fire caps, badges, signals, or devices, adopted or used by the Fire Department, shall be deemed guilty of a misdemeanor, and shall be subject to a fine of not less than twenty-five dollars, or more than two hundred and fifty dollars and by imprisonment for a term not less than ten days, or more than three months; and all fines so collected, to be paid over to the Trustees of the New York Fire Department Fund.

§ 20 This act shall take effect on the second Tuesday in May next.

RESOLUTION AS TO DIES AND BADGES.

Resolved, That the Commissioner of Repairs and Supplies be, and he is hereby directed to advertise for proposals for two dies, and a sufficient number of badges, of the size, form, and material specified, for the use of the Fire Department, as adopted by the committees appointed for that purpose, and submit the contract for the same to the Common Council for confirmation.

Adopted by the Board of Councilmen, May 21, 1855
Adopted by the Board of Alderman, May 23, 1855
Approved by the Mayor, May 25, 1855

EXCERPT FROM

Annual Report of the Board of Trustees of the Fire Department
December 30, 1855

"The law requiring every fireman, when on duty, to wear either the badge or a fire-cap, and prohibiting any but firemen from wearing them, went into operation in the early part of the present month. The intention of the law is to keep all persons but firemen from the vicinity of fires, especially thieves and other bad characters; for this the law gives the police absolute power of arrest and expulsion, and it requires their active co-operation to carry it out; but it is to be regretted that the police have not given that co-operation which the Department has a right to expect, and the orders of the Mayor, to that effect, are in a great measure disregarded. The members of this Department will cheerfully sustain the police in the discharge of their duty; and it is to be hoped that if the latter fail, the former will not be held responsible for the acts of such thieves and vagabonds as the police may hereafter permit to prowl around the neighborhood of fires.

I would advise an amendment to the law - that the badges be abolished altogether, save for the Engineer Board and officers of companies."

Harry Howard, Chief Engineer

EXCERPT FROM

1858 Annual Report of the New York Fire Department

"I find the law authorizing the issue of badges to members of the Department to be productive of much evil. It has not accomplished what was designed by it introduction, but has added largely to the cause for complaint it was designed to remedy. The number of badges that are either lost or stolen amounts to over one thousand at the present time, and it is growing to be a source of dissatisfaction among members of the department at the necessity of wearing them. The ownership of a stolen badge gives free access to thieves, and other dangerous persons, to the close proximity of fires, where, under the recognized shield of the Fire Department they can perform their nefarious calling without suspicion or interruption. It is a well-known fact that coats, on which badges are worn, have been stolen from different apparatus while on their way to fires; and it is not difficult to conjecture that they were taken solely for the purpose of procuring badges to give recognized admittance to a building on fire.

I would advise an amendment to the law - that the badges be abolished altogether, save for the Engineer Board and officers of companies."

Harry Howard, Chief Engineer

AN ORDINANCE

The Mayor, Alderman, and Commonalty of the City of New York in Common Council convened, do ordain as follows

§ 1. All members of the New York Fire Department, and such exempt members as are authorized by the Commissioners of the Fire Department, shall hereafter when on duty as firemen, wear the leathern cap, as heretofore used, or a badge, as hereinafter provided.

§ 2. The badge mentioned in the preceding section shall be made of Prince's metal, bearing the words "New York Fire Department" and each badge shall bear a distinct number in raised figures thereon, of white metal; and the badge to be worn by exempt firemen, as provided for in Section 3d., of this ordinance, shall be composed of white metal, with figures thereon of Prince's metal; but in all other respects to conform to the badge herein described for use of the active member of the Department; said badges shall be struck from dies, which shall be exclusively the property of the Corporation of the Clerk of the Common Council; and all of such badges shall be numbered, as the Commissioner of the New York Fire Department may direct.

§ 3. The said badges shall be deposited with the Commissioners of the New York Fire Department, who shall have sole charge of the distribution of the same; and it shall be the duty of the said Commissioners to keep a register of the names of all persons who now are, or who may hereafter become members of the New York Fire Department, and of the number of the company to which said persons are attached; and also of the names of such exempt firemen as may, in writing, be permitted by the said Commissioners to wear the badges aforesaid, and of the number of the badge delivered to each of said firemen and exempt firemen.

§ 4. It shall be the duty of the Commissioners of the New York Fire Department to deliver to each fireman one of the badges aforesaid; and the said Commissioners shall also deliver one of the aforesaid badges to each of such exempt firemen as may, by virtue of section 16, of an act of the Legislature of the State of New York, entitled, "An act for the better regulation of the firemen in the city of New York," passed March 29th, 1855, be permitted by said Commissioners to wear the same.

§ 5. The badges herein named shall be the exclusive property of the city of New York, and when any member of the Fire Department shall resign, or be suspended or expelled therefrom, it shall be the duty of the Foreman and the secretary of the company to which such member is attached to make a return of such suspension, expulsion, or resignation, forthwith, to the Chief Engineer, together with the badge formerly worn by such member, and the Chief Engineer shall report weekly, to the Commissioners aforesaid, the names of such members of the Fire Department as shall have resigned, or been suspended or expelled therefrom, since his last report, and shall with said report return to the said Commissioners the badges worn by said members.

§ 6. If the foreman and secretary of any company or either of them, shall fail to comply with the provisions of the preceding section in relation to the return of badges, no return of members elected by such company shall be received therefrom, unless good and satisfactory cause shall be shown to the Commissioners why said badges are not returned; and should any member of the Fire Department lose his badge, it shall be his duty to report said loss within two weeks to the said Commissioners, and it shall be their duty to inquire into the circumstances of the loss within two weeks to said Commissioners, and it shall be their duty to inquire into the circumstances of the case, and unless they are satisfied that such loss was without fault on the part of said member, they shall have power to suspend or expel said member from the Department, in their discretion; but should the loss be satisfactorily accounted for by the said Commissioner, then they shall grant a new badge to said member, who shall pay one dollar, upon receipt of the same, as a penalty for said loss (which money shall be paid to the Treasurer of the Fire Department Fund by said Commissioners they taking his receipt for the same), and any member of the Department not complying with the above provision shall be expelled.

§ 7. Every exempt fireman receiving a badge shall have the permission to wear the same renewed some time during the month of May in each year, otherwise such permission shall be deemed revoked, and the said Commissioners may, at any time, revoke any permission so granted by them; and in case such permission is not renewed, or is revoked, it shall be the duty of such exempt fireman to restore the badge previously worn by him, to the said Commissioners.

§ 8. Any member of the New York Fire Department, or exempt fireman, who shall loan his badge to any person, shall be liable to the penalties as set forth in section 6 of this ordinance; and if any badge shall become broken of the figures be displaced while n the possession of said member or exempt fireman, he shall return the badge forthwith to the said Commissioners, and in default of the same, said member may be expelled or suspended, and such exempt fireman shall have the permission to wear the same revoked.

§ 9. Any member of the New York Fire Department, or exempt fireman, who shall violate any of the forgoing sections of this ordinance, shall be subject to a fine of not less than twenty-five, nor over two hundred and fifty dollars, and to imprisonment for the term of ten days; and all fines so collected shall be paid over to the Treasurer of the New York Fire Department Fund.

§ 10. It shall be the duty of the police, when a fire occurs, to form a line, at least two hundred feet distant from the said fire, on each side thereof; and they shall not, under any circumstances, permit any person to pass such line, unless said person shall wear the uniform of the Insurance Patrol, or be a member of the Common Council, a member of the Police Department, an owner of property or resident in the prescribed lines.

§ 11. It shall be the duty of the police to promptly remove from within the said lines, all persons not designated in the last preceding section, and all persons refusing to retire at the request of a policeman, from within the said lines, shall be deemed guilty of a misdemeanor and forthwith arrested.

§ 12. All ordinances, and parts of ordinances, now existing, so far as the same are inconsistent with any section of the provisions of this ordinance, are hereby repealed.

§ 13. This ordinance shall take effect immediately.

Adopted by the Board of Councilmen, June 8, 1855
Adopted by the Board of Aldermen, June 13, 1855
Adopted by the Mayor, June 14, 1855

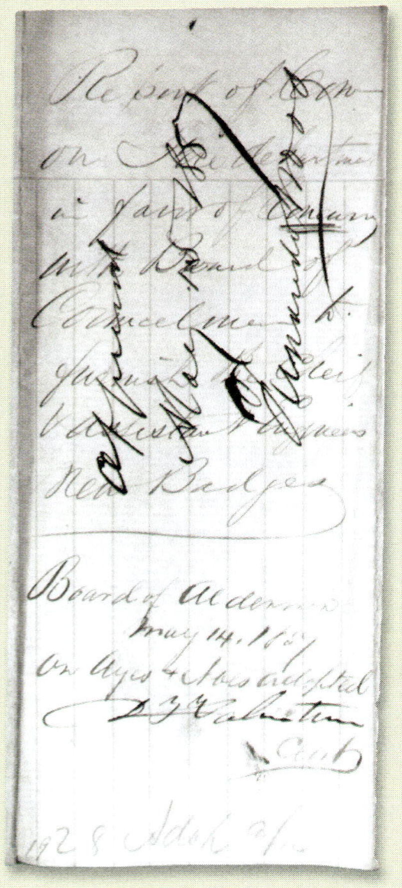
Report of Com on Fire Dept in favor of concurring with Board of Councilmen to furnish Chief & Assistant Engineers New Badges

Approved May [illegible] 1857

Board of Aldermen May 14. 1857 On Ayes & Noes adopted D. T. Valentine Clerk

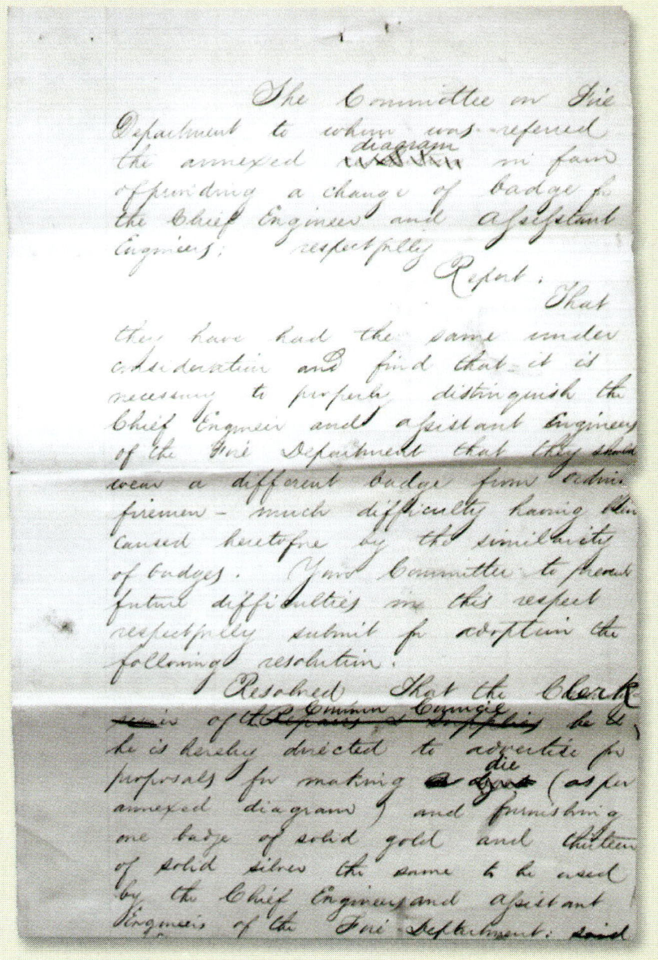
The Committee on Fire Department to whom was referred the annexed diagram in favor of providing a change of badge for the Chief Engineer and Assistant Engineers; respectfully Report. That they have had the same under consideration and find that it is necessary to properly distinguish the Chief Engineer and Assistant Engineers of the Fire Department that they should wear a different badge from ordinary firemen – much difficulty having been caused heretofore by the similarity of badges. Your Committee to prevent future difficulties in this respect respectfully submit for adoption the following resolution:

Resolved, That the Clerk of the Common Council be & he is hereby directed to advertise for proposals for making a die (as per annexed diagram) and furnishing one badge of solid gold and thirteen of solid silver the same to be used by the Chief Engineer and Assistant Engineers of the Fire Department: said

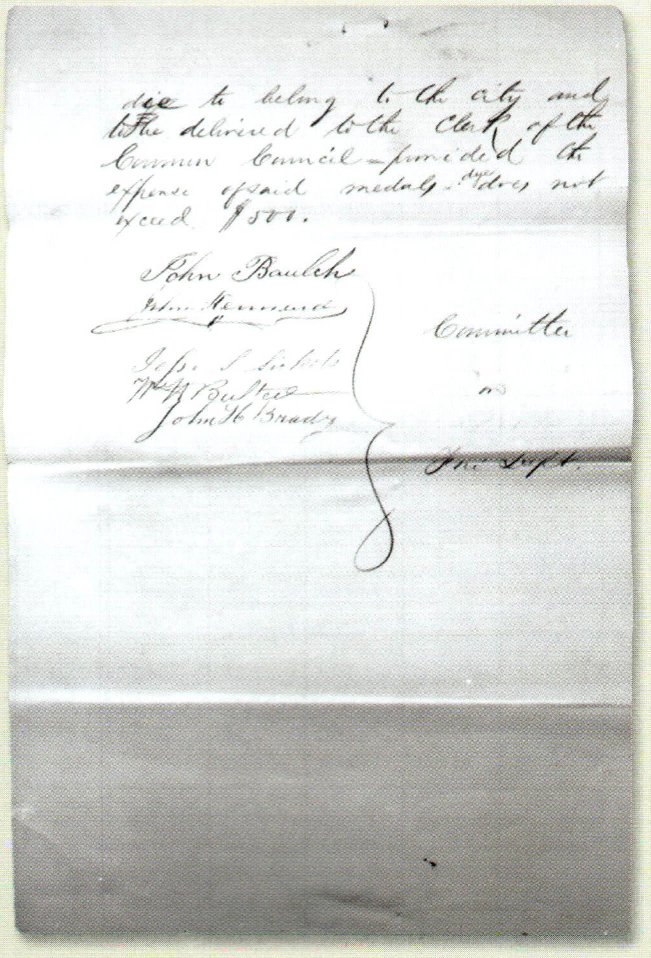
die to belong to the city and to be delivered to the Clerk of the Common Council – provided the expense of said medals & dies does not exceed $500.

John Baulch
John [illegible]
Jos. [illegible]
Wm. [illegible]
John H. Brady

Committee on Fire Dept.

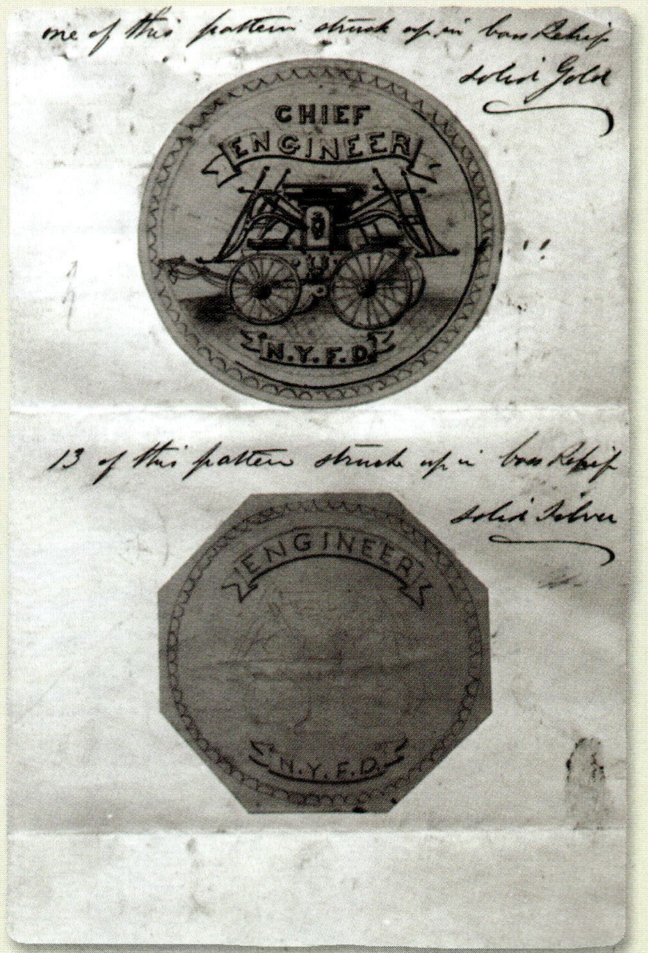
one of this pattern struck up in bas relief solid Gold

13 of this pattern struck up in bas relief solid Silver

Report of the Committee on the Fire Department in favor of concurring with Board of Councilmen to furnish the Chief and Assistant Engineers new badges.

**RESOLUTION OF THE
BOARD OF COMMISSIONERS
NEW YORK FIRE DEPARTMENT**

RESOLVED, That each and every exempt member of the New York Fire Department to whom permission shall be given by this Board to use a Badge, shall on any change of residence of such exempt Member give immediate notice of the same to this Board under penalty of having the said permission revoked.

Date unknown, circa 1855

AN ORDINANCE

The Mayor, Aldermen, and Commonalty of the City of New York in Common Council convened, do ordain as follows.

Resolved, That the Clerk of the Common Council be, and he is hereby directed to advertise for proposals for making a die (as per annexed diagram) and furnishing one badge of solid gold, and thirteen solid silver, the same to be used by the Chief Engineer and Assistant Engineers of the Fire Department; said die to belong to the city, and to be delivered to the Clerk of the Common Council, provided the expenses of said medals and dies does not exceed five hundred dollars.

Adopted by the Board of Councilmen, March 10, 1857
Adopted by the Board of Aldermen, May 14, 1857
Adopted by the Mayor, May 15, 1857

EXCERPT FROM

AN ORDINANCE to Establish a Fire Department, &c in the City of Brooklyn.

The Common Council of the City of Brooklyn do ordain as follows:

Article I.

§16. The members of the department shall, when on duty, be distinguished by their badge, or as follows, respectively. Every engineer shall wear a white fire cap, on the front of which shall be painted the initials of his name, and the word chief engineer or assistant engineer, as his office may be. Every foreman and assistant foreman of a company shall wear a fire cap painted black, with a white front, on which shall be the initials of his name, the number of his company, and the word foreman or assistant, as his office shall be. Every other fireman shall wear a fire cap painted black, on the front of which, in white lettering, shall be the initials of his name and the number of his company, as Engine No. 1, Hose No. 2, as the case may be.

§21. The badge to be worn by the firemen of the Western District shall be made of princes metal, with a distinct number in raised figure thereon of white metal, each bearing the words Brooklyn Fire Department, with the letters W.D. to designate the district, and the badge to be worn by exempt firemen of the district shall be made of white metal, with yellow raised numbers, but in all other respects to conform to the badge herein described for use by active members of the department.

§22. The said badges shall be deposited with the Commissioners of the Fire Department of the Western District, who shall have sole charge of the distribution of the same, and it shall be the duty of said board of fire commissioners to keep a register of the names of all persons who now are or who may hereafter become members of the said department, and of the company to which said persons are attached, and also of the names of such exempt firemen as may in writing be permitted by the said commissioners to wear the badge aforesaid, and of the number of the badge delivered to each of said firemen and exempt firemen.

§23. And it shall be the duty of said commissioners to deliver to each and every member of the department who is approved by them, one of the aforesaid badges of princes metal and to each such exempt fireman as they may consider entitled thereto, by producing proper evidence of their term of service having expired, and of the aforesaid badges of white metal.

§24. The badges herein mentioned shall be the exclusive property of the city of Brooklyn, and when any member of the department shall resign or be removed therefrom, he shall surrender his badge to the foreman or secretary of the company to which such member was attached, and it shall be the duty of said foreman and secretary to make a return of such removal or resignation forthwith to the chief engineer, together with the badge formerly worn by such member, and the chief engineer shall weekly report to the commissioners aforesaid the names of such members of the fire department as shall have resigned or been removed therefrom since his last report, and shall, with said report, return to said commissioners the badges worn by such members.

§26. Every exempt fireman to whom permission shall be given by the said commissioners to wear the aforesaid badge, shall have the said permission renewed some time during the month of May in each year, otherwise such permission shall be deemed revoked; and the said commissioners may at any time revoke any permission so granted by them; but in case such permission is not renewed or is revoked, it shall be the duty of such exempt fireman to restore the badge previously worn by him to the said commissioners.

§27. It shall be the duty of the police when a fire occurs, to form a line at least two hundred feet distant from the said fire, on either side thereof; and they shall not, under any circumstances, permit any person to pass said line, unless said person shall wear the uniform or fire department badge, or be a member of the common council, a member of the police department, or owner of property within, or a resident in the prescribed lines, or an insurance agent; provided, however, that nothing herein contained shall be so construed as to prevent the chief engineer, or, in his absence, an assistant engineer, from calling upon the citizens to assist in extinguishing the fire.

§29. Any person refusing to deliver up his badge upon the requisition of the board of fire commissioners of the Western District, shall be subject to a fine of not less than twenty-five dollars, and to imprisonment for a term of not less than ten days.

Adopted in Common Council: November 15th, 1858, and by unanimous consent ordered to take effect immediately.

Approved by the Mayor, Nov. 20, 1858

AN ORDINANCE

To amend an Ordinance for the better Regulation of the firemen of the city of New York, passed June 15th, 1855

§ 1. All members of the New York Fire Department, and such exempt members as are authorized by the Commissioners of the Fire Department, shall hereafter when on duty as firemen, or at fires, wear the leathern cap, as heretofore used, or a badge, as hereinafter provided, which said badge shall be worn in a plain, conspicuous manner, on the breast, and shall be worn and visible during the whole time that such member or, exempt member, shall remain at a fire; and any firemen neglecting or refusing to wear his badge, as above specified, shall be subject to suspension or expulsion for the Department; and any exempt fireman so refusing or neglecting shall be subject to a fine of not less than twenty-five, nor over two handred and fifty dollars, and imprisonment not less than ten days; and all fines that are collected shall be paid over to the Treasurer of the Fire Department Fund.

§ 2. The badge mentioned in the preceding section shall be made of Prince's metal, bearing the letters "N.Y.F.D." and each badge shall bear a distinct number in raised figures thereon, of white metal, the same to represent the number of each company, and also a registered number, in small figures, stamped thereon; and the badge to be worn by exempt firemen, as provided for in Section 3d., of this ordinance, shall be composed of white metal, with figures thereon of Prince's metal; but in all other respects to conform to the badge herein described for use of the active member of the Department; said badges shall be struck from dies, which shall be exclusively the property of the Corporation of the Clerk of the Common Council; and all of such badges shall be numbered, as the Commissioner of the New York Fire Department may direct.

§ 3. The said badges shall be deposited with the Commissioners of the New York Fire Department, who shall have sole charge of the distribution of the same; and it shall be the duty of the said Commissioners to keep a register of the names of all persons who now are, or who may hereafter become members of the New York Fire Department, and of the number of the company to which said persons are attached; and also of the names of such exempt firemen as may, in writing, be permitted by the said Commissioners to wear the badges aforesaid, and of the number of the badge delivered to each of said firemen and exempt firemen.

§ 4. It shall be the duty of the Commissioners of the New York Fire Department to deliver to each fireman one of the badges aforesaid; and the said Commissioners shall also deliver one of the aforesaid badges to each of such exempt firemen as may, by virtue of section 16, of an act of the Legislature of the State of New York, entitled, "An act for the better regulation of the firemen in the city of New York," passed March 29th, 1855, be permitted by said Commissioners to wear the same.

§ 5. The badges herein named shall be the exclusive property of the city of New York, and when any member of the Fire Department shall resign, or be suspended or expelled therefrom, it shall be the duty of the Foreman and the secretary of the company to which such member is attached to make a return of such suspension, expulsion, or resignation, forthwith, to the Chief Engineer, together with the badge formerly worn by such member, and the Chief Engineer shall report weekly, to the Commissioners aforesaid, the names of such members of the Fire Department as shall have resigned, or been suspended or expelled therefrom, since his last report, and shall with said report return to the said Commissioners the badges worn by said members.

§ 6. If the foreman and secretary of any company or either of them, shall fail to comply with the provisions of the preceding section in relation to the return of badges, no return of members elected by such company shall be received therefrom, unless good and satisfactory cause shall be shown to the Commissioners why said badges are not returned; and should any member of the Fire Department lose his badge, it shall be his duty to report said loss within two weeks to the said Commissioners, and it shall be their duty to inquire into the circumstances of the loss within two weeks to said Commissioners, and it shall be their duty to inquire into the circumstances of the case, and unless they are satisfied that such loss was without fault on the part of said member, they shall have power to suspend or expel said member from the Department, in their discretion; but should the loss be satisfactorily accounted for by the said Commissioner, then they shall grant a new badge to said member, who shall pay one dollar, upon receipt of the same, as a penalty for said loss (which money shall be paid to the Treasurer of the Fire Department Fund by said Commissioners, they taking his receipt for the same), and any member of the Department not complying with the above provision shall be expelled.

§ 7. Every exempt fireman receiving a badge shall have the permission to wear the same renewed some time during the month of May in each year, otherwise such permission shall be deemed revoked, and the said Commissioners may, at any time, revoke any permission so granted by them; and in case such permission is not renewed, or is revoked, it shall be the duty of such exempt fireman to restore the badge previously worn by him, to the said Commissioners.

§ 8. Any member of the New York Fire Department, or exempt fireman, who shall loan his badge to any person, shall be liable to the penalties as set forth in section 6 of this ordinance; and if any badge shall become broken of the figures be displaced while n the possession of said member or exempt fireman, he shall return the badge forthwith to the said Commissioners, and in default of the same, said member may be expelled or suspended, and such exempt fireman shall have the permission to wear the same revoked.

§ 9. Any member of the New York Fire Department, or exempt fireman, who shall violate any of the forgoing sections of this ordinance, shall be subject to a fine of not less than twenty-five, nor over two hundred and fifty dollars, and to imprisonment for the term of ten days; and all fines so collected shall be paid over to the Treasurer of the New York Fire Department Fund.

§ 10. It shall be the duty of the police, when a fire occurs, to form a line, at least two hundred feet distant from the said fire, on each side thereof; and they shall not, under any circumstances, permit any person to pass such line, unless said person shall wear the uniform of the Insurance Patrol, or be a member of the Common Council, a member of the Police Department, an owner of property within or resident in the prescribed lines.

§ 11. It shall be the duty of the police to promptly remove from within the said lines, all persons not designated in the last preceding section, and all persons refusing to retire at the request of a policeman, from within the said lines, shall be deemed guilty of a misdemeanor and forthwith arrested.

§ 12. All ordinances, and parts of ordinances, now existing, so far as the same are inconsistent with any section of the provisions of this ordinance, are hereby repealed.

§ 13. This ordinance shall take effect immediately.

Adopted by the Board of Councilmen, May 30, 1859
Adopted by the Board of Aldermen, June 13, 1859
Adopted by the Mayor, June 27, 1859

EXCERPTS FROM

Rules and Regulations

OF THE

METROPOLITAN FIRE DEPARTMENT.

•••••••••••••••••••••

UNIFORM.

§ 2. Uniform of the officers and members of the Metropolitan Fire Department:

FATIGUE CAP.

FOR CHIEF ENGINEER. — To be made of dark blue cloth, same as the uniform coat of the United States Navy pattern, with a plain solid black patent leather visor two inches wide, with rounded corners; top seven inches wide and seven and one-half to eight inches long; band one and one-half inches wide' quarters one and three-quarter inches high; lining of red leather, to be sewed into the seam of top and quarters; a narrow welt around the bottom of band and also around the top seam of cap; inner band to be of strong, serviceable leather; chin strap of black patent leather, one-half inch wide, with white metal slide worn in the center of the front; one small regulation button on each side of cap; insignia of office to be gold embroidered star, on dark blue cloth, of five trumpets joined at the mouth-pieces, measuring two inches from point to point, placed in the center of the front, the lower points resting upon the chin strap.

FOR ASSISTANT ENGINEER. — Same as for Chief Engineer, except the insignia of office, which will be three cross trumpets embroidered in gold, measuring one and three-quarter inches from mouth-piece to bell, placed in the center of the cap front, bells outward and resting upon upper edge of band.

FOR DISTRICT ENGINEERS. — Same as for Assistant Engineer, except the insignia of office, which will be two cross trumpets embroidered in gold, measuring one and three-quarter inches from mouth-piece to bell.

FOR FOREMEN OF ENGINE COMPANIES. — Same as for District Engineers, except the insignia of office, which will be two parallel trumpets, one and three-quarter inches long, with the number of the company in block figures, one-half inch long, all embroidered in gold upon a circle of light blue cloth, the trumpets placed vertically, bells downward, in the center of the cap front, the bells resting upon the upper edge of cap band.

FOR ASSISTANT FOREMEN OF ENGINE COMPANIES. — Same as for Foremen of Engine Companies, except the insignia of office, which will be one trumpet, one and three-quarter inches long, placed horizontally with the number of the company in block figures, one-half inch long, above the trumpet.

FOR FOREMEN OF LADDER COMPANIES. — Same as for Foremen of Engine Companies, except that the embroidery will be upon scarlet cloth instead of light blue.

FOR ASSISTANT FOREMEN OF LADDER COMPANIES. — Same as for Assistant Foremen of Engine Companies, except that the embroidery will be upon scarlet cloth instead of light blue.

FOR ALL FIREMEN. — Same as for company officers, except the device, which will be a white metal Maltese cross, one and three-quarter inches square, with the letters M.F.D. and the department number of the wearer on the arms, placed in the center of the front of the cap, the lower arm touching the cap band.

FOR HOSEMEN AND LADDERMEN. — Same as for firemen, omitting the department number on the device.

Officers and men will wear, in stormy weather, a cover to the fatigue cap, made of oiled silk. The cap, device, and cap-cover, all to be according to the pattern at headquarters of the Department.

PERSONS TO BE ADMITTED TO FIRES OTHER THAN MEMBERS OF THE FORCE.

§ 31. The Fire Marshal and Assistant Fire Marshal, Superintendent and Inspectors of Unsafe Buildings, Inspectors of Gas, reporters of the press, and the officers of each fire insurance company. They shall be designated by a badge to be worn on the outside of the coat, on the left breast, of the following description, of yellow metal, in the form of a shield, surmounted with fire cap and crossed pipes; in the center of the badge, within a shield, a design representing the seal of the Department; over this, in raised letters, the words "Metropolitan Fire Department," and below, large figures of white metal, designating the number of the badge. Officers and members of the Insurance Patrol shall be designated by a red fire cap, with number and initials of wearer in front; or a red leather hat-band, with the letters "I.P." and the number in white or gilt letters in front thereof.

Board of Commissioners.

Alexander Shaler, *President.*
Joshua G. Abbe, *Treasurer.*

J. Bailey Meyers — Monmouth B. Wilson.
James Galway. — Charles E. Gildersleeve, *Secretary.*

Committees.

APPOINTMENTS. Shaler, Abbe, Meyers, Galway, and Wilson.	FINANCE. Meyers and Shaler.
DISCIPLINE. Shaler, Meyers, and Wilson.	TELEGRAPH. Meyers and Abbe.
SUPPLIES. Abbe, Shaler, and Wilson.	BUILDINGS. Galway and Meyers.
APPARATUS. Wilson and Galway.	HOSE. Galway and Wilson.

EXCERPTS FROM

Metropolitan Fire Department,

OFFICE CHIEF ENGINEER,
FIREMEN'S HALL, NEW YORK, JANUARY 16, 1868.

GENERAL ORDERS No. 1.

I. Pursuant to a resolution adopted by the Board of Commissioners on the 15th inst., the following Bill of Dress is hereby promulgated for the information and government of this Department:

FATIGUE CAP.

For Chief Engineer.—To be made of dark blue cloth same as the uniform coat of the U. S. Navy pattern, with a plain solid black patent leather visor two inches wide, with rounded corners; top seven inches wide and seven and one half to eight inches long; band one and one-half inches wide; quarters one and three-quarter inches high; lining of red leather, to be sewed into the seam of top and quarters; a narrow welt around the bottom of band and also around the top seam of cap; inner band to be of strong serviceable leather; chin strap of black patent leather, one-half inch wide, with white metal slide worn in the centre of the front; one small regulation button on each side of cap; insignia of office to be a gold embroidered star, on dark blue cloth, of five trumpets joined at the mouth-pieces, measuring two inches from point to point, placed in the centre of the front, the lower points resting upon the chin strap.

For Assistant Engineer.—Same as for Chief Engineer, except the insignia of office, which will be three cross-trumpets embroidered in gold, measuring one and three-quarter inches from mouth-piece to bell, placed in the centre of the cap front, bells downward and resting upon upper edge of band.

For District Engineers.—Same as for Assistant Engineer, except the insignia of office, which will be two cross-trumpets embroidered in gold, measuring one and three-quarter inches from mouth-piece to bell.

For Foremen of Engine Companies.—Same as for District Engineers, except the insignia of office, which will be two parallel trumpets, one and three-quarter inches long, with the number of the company in block figures, one-half inch long, all embroidered in gold upon a circle of light blue cloth, the trumpets placed vertically, bells downward, in the centre of cap front, the bells resting upon the upper edge of cap band.

For Assistant Foremen of Engine Companies.—Same as for foremen of Engine Companies, except the insignia of office, which will be one trumpet, one and three-quarter inches long, placed horizontally, bell to the right, with the number of the company in figures one-half inch long, above the trumpet.

For Foremen of Hook and Ladder Companies.—Same as for foremen of Engine Companies, except that the embroidery will be upon scarlet cloth instead of light blue.

For Assistant Foremen of Hook and Ladder Companies.—Same as for assistant foremen of Engine Companies, except that the embroidery will be upon scarlet cloth instead of light blue.

For all Firemen.—Same as for company officers, except the device, which will be a white metal Maltese cross, one and three-quarter inches square, with the letters M. F. D. and the Department number of the wearer on the arms, placed in the centre of the front of the cap, the lower arm touching the cap band.

For Hosemen and Laddermen.—Same as for firemen, omitting the Department number on the device.

Officers and men will wear, in stormy weather, a cover to the fatigue-cap, made of oiled silk. The cap, device, and cap cover, all to be according to the pattern at Headquarters of the Department.

INSIGNIA OF OFFICE.

For Chief Engineer.—A star formed of five trumpets joined at the mouth-pieces, measuring one and a half inches from point to point, embroidered in gold upon a circle of dark blue cloth, to be worn upon the front corners of the collar of the uniform coat.

For Assistant Engineer.—Three cross-trumpets, bells downward, measuring one and a half inches from mouth-piece to bell, embroidered in gold, upon a circle of dark blue cloth, to be worn same as Chief Engineer, bells touching outer edge of collar.

For District Engineer.—Two cross-trumpets, bells downward, measuring one and a half inches in length, embroidered in gold upon an oval of dark blue cloth, to be worn same as Assistant Engineer.

For Foremen.—Two parallel trumpets, bells downward, measuring one and a half inches, embroidered in gold upon an oval of light blue cloth for foremen of Engine Companies, and of scarlet cloth for foremen of Hook and Ladder Companies, to be worn same as Assistant Engineers.

For Assistant Foremen.—One trumpet, measuring one and a half inches, embroidered in gold upon an oval of light blue cloth for assistant foremen of Engine Companies, and of scarlet for assistant foremen of Hook and Ladder Companies, to be worn same as Assistant Engineer.

II. R. V. Mackey, District Engineer, is hereby designated as Inspecting Officer.

III. All uniforms and parts of uniforms, made after this date, will conform strictly to the prescribed Bill of Dress, and receive an inspection before being worn; and no insignia of office except the one prescribed will be worn upon any part of the uniform. Officers are expected to conform to the changes prescribed with as little delay as possible, reporting to the Inspecting Officer as soon as fully uniformed. Company Commanders will see that the provisions of Circular No. 4, from the Office of Board of Commissioners, Dec. 7th, 1867, are complied with on the part of their men.

IV.—For the inspection of new garments or alterations to old ones, Engineer Mackey will be in attendance at this Office at 12 M. daily. The General Inspection referred to in Circular No. 4 will take place at a day hereafter to be designated.

By order of

ELISHA KINGSLAND,
Chief Engineer.

OFFICIAL.

____
District Engineer.

Metropolitan Fire Department,

OFFICE CHIEF ENGINEER,

FIREMEN'S HALL, NEW YORK, JULY 2, 1868.

GENERAL ORDERS No. 15.

By direction of the Board of Commissioners, the following additions to General Orders No. 1, C. S., Office of Chief Engineer, are made to the paragraphs under the head of "Fatigue Cap."

For Summer wear.—From the 15th of June until the 15th of September, officers will wear a Panama, and the men a straw hat of the style and color of the pattern at Headquarters of the Department, with a black silk band $1\frac{3}{8}$ inch wide, tied with a bow on the left side.

Officers will wear the insignia of office now worn upon the fatigue cap sewed upon the front of the hat. The men will wear their Department numbers in block figures, of white metal, $\frac{3}{4}$ inch long, set upon a piece of patent leather $1\frac{1}{2}$ inch wide and $2\frac{1}{2}$ inches long, securely sewed upon the front of the hat, the lower edge touching the brim.

The color of the leather upon which the figures are set will be as follows:

For men of Engine Companies, Light Blue.

For men of Hook and Ladder Companies, Scarlet.

By order of

ELISHA KINGSLAND,

Chief Engineer.

OFFICIAL.

Thomas Sullivan

District Engineer.

EXCERPTS FROM

OFFICE BOARD OF COMMISSIONERS
NEW YORK, April 2, 1879

GENERAL ORDERS NO. I.

I. — The following amended Bill of Dress for the Uniformed Force of this Department, having been adopted, is hereby promulgated for its information and guidance:

HATS FOR SUMMER WEAR.

For all Officers. — Of white Mackinac braided straw (braids about one-half an inch wide), sides slightly tapering to crown and about four and one-half inches high, with black ribbed silk ribbon band seven-eighths of an inch wide, tied in a single bow knot on left side; brim about two and one-half inches wide, bound with seven-eighths inch black ribbed silk ribbon; the sweat band to be of leather about two inches wide, and the inside of the hat to be lined with gauze.

For all other Members. — Of brown Mackinac braided straw (braids about five-eighths of an inch wide), side about three and one-half inches high, slightly tapering to crown, with black ribbed silk ribbon band one and three-quarters of an inch wide, tied in a single bow knot on left side; brim about two and one-quarter inches wide, bound with seven-eighths inch black ribbed silk ribbon; the sweat band to be of leather about one and one-half inches wide; the bottom of crown to be lined with paper. The registered number in figures of German silver five-eighths of an inch long, fastened on a strip of black patent leather, to be sewed on the front of the band.

INSIGNIA OF RANK.

Chief of Department. — Five trumpets measuring one and five sixteenths inches each, crossed with bells outward and projecting beyond mouth pieces so as to form a design one and one-half inches in diameter, to be embroidered in gold upon a circle of dark blue cloth.

For Assistant Chief of Department. — Same as for Chief of Department, except that there shall be three trumpets measuring one and one-half inches each, crossed with bells downward.

For Chiefs of Battalion. — Same as for Chief of Department, except that there shall be two trumpets, embroidered on an oval of dark blue cloth.

For Foremen. — Same as for Chiefs of Battalion, except that the trumpets are to be parallel, with bells downward and close together, embroidered on an oval of cloth, dark blue for engine, scarlet for hook and ladder, and light green for chemical engine companies. The insignia for the cap is to have the number of the company embroidered in gold midway between the trumpets, which are to be placed three-eighths of an inch between bells.

For Assistant Foremen. Same as for Foremen, except that there will be only one trumpet. The insignia for the cap to have the trumpet placed horizontally, the bell to the left and the number of the company embroidered in gold above the centre.

PATROL BADGE.

For Engineers and Assistant Engineers of Steamer. — A white metal maltese cross two inches square, with the letters "F.D." on the top arm, the letters "N.Y." on the lower arm, a hook and ladder crossed on the right arm, and a hydrant on the left arm, and with the company number in figures three-eighths of an inch long on a black ground, in a circle one inch in diameter in the centre. The whole badge to be two inches from point to point of arm, and to be fastened with a pin and improved catch.

For all other Members. — Same as for Engineers and Assistant Engineers of Steamer, except that the registered number of each member is to be substituted for the company number in the centre.

Headquarters Fire Department City of New York.
OFFICE BOARD OF COMMISSIONERS

CORNELIUS VAN COTT,
HENRY D. PURRY, } Commissioners
RICHARD CROKER,

GENERAL ORDERS NO. 4. }

New York, May 20, 1884.

I. — It having been found that the badges issued for admission within fire lines only were used for a variety of unauthorized purposes, the holders were notified to surrender them on or before April 20, 1884. A new badge was provided, and the Police Department, at the request of the Board, issued the following instructions:

OFFICE OF THE SUPERINTENDENT OF POLICE
OF THE CITY OF NEW YORK,
No. 300 Mulberry Street.

NEW YORK, April 18, 1884

General Order No. 261.
Captain...
....*Precinct.*

On and after May 1st, 1884, you will exclude from within the Police lines at fires, any and all persons who do not prominently display a badge exactly like the design herewith, except the Mayor, Alderman, the heads of all the Municipal Departments, the uniformed firemen and insurance patrolmen.

You will also, on and after May 1, 1884, take any and all badges formerly issued by the Fire Department from all persons attempting to make use of or displaying them, and send such badges to the Central Office. Attention is called to the following provisions of law:

CHAPTER 410 (CONSOLIDATION ACT), LAWS OF 1882

SEC. 443. It shall be the duty of the Fire Department and of the Police Board, their respective officers and men, to cooperate together in all proper ways, and the said board and department may respectively provide for protection against fire, and for the arrest to commit any crime against the laws of this State, ***or violate any rule or regulation of said Board or Department.***

SEC. 451. During the actual prevalence of any fire, it shall and may be lawful for the officers of the Police and Fire Departments to remove or cause to be removed and kept away from the vicinity of such fire all idle and suspicious persons, ***and all persons not fit to be employed, or not actually and usefully employed, in their judgement, in aiding the extinguishment of such fire, or in the preservation of property in the vicinity thereof.***

You will instruct the members of your command in the duties required of them under this order, so that its provisions may be strictly enforced.

The framed copy of this order you will cause to be hung up in the sitting-room of your station-house.

THOMAS W. THORNE,
Acting Superintendent.

....
Inspector...............................*District.*

II. — All officers, members and employees of this Department are hereby directed to aid in the enforcement of the above order, by calling the attention of the Police force to all infractions of its provisions.

III. — The new badge is, as is distinctly stated upon its face, to "admit within fire lines," and cannot legitimately be used to admit to buildings or premises whether in or outside of the fire lines, or for any other purpose by any one than the purpose for which it was issued.

IV. — The badges are issued upon payment of a deposit of two dollars for each one, as follows:

To each local fire insurance company not more than three.

To each principal agency in this city of fire insurance companies located outside of the city, whether it be an agency of one or more companies, not more than two.

To each newspaper published in the city, or press association located therein, such a number as the Board may determine.

To such public officials and others as the Board may direct.

V. — The badges will also be issued, without deposit, to the officers of the uniformed force; and to other subordinates of the Department not in the uniformed force, without deposit, upon their application granted by the Board; in both cases on condition that the penalty for losing the badge shall be a fine of five dollars.

VI. — The Secretary is charged with carrying out the rules relating to the badges above referred to, and will render an account of the same to the Board, and of the deposits to the Treasurer, quarterly, on or before the tenth day of January, April, July and October of each year.

VII. — For the use of officers and members of the uniformed force detailed for duty in theatres, a badge of the accompanying design was adopted, which shall be known and

designated as the Theatre Detail Badge, and be issued and used subject to the following rules:

1st. The Property Record Clerk will have charge of the badges, account for the same by their numbers, and issue them to the Assistant Chief of Department and Second Assistant Chief of Department upon their demand, in such numbers from time to time, as may be necessary; but all badges in excess of the regular number required and issued to them for extraordinary occasions, shall be returned to the Property Record Clerk without delay.

2d. The Assistant Chiefs of Department shall issue to company commanders the number required by them for details to be furnished, keeping a record of their numbers.

3d. The company commanders shall keep the badges so issued to them, in their own possession, until required for use by the officers or men detailed for this duty, who shall return them immediately upon reporting back therefrom. The number of each badge so furnished to details shall be recorded in the company journal, both at the time of leaving and returning to quarters.

4th. The badge shall be substituted for the coat badge during service on theatre detail and shall be worn on the left breast at the level of the second button from the top of the coat and midway between the centre of the body and the side.

5th. The use of the badge by any one except officers and members of the uniformed force, detailed for theatre duty, and in uniform, is prohibited.

6th. On the reports of duty performed by officers and members detailed for theatre duty, as prescribed by General Order No. 24, O.B.C., series of 1881, they shall also report by name or number, all officers or members of the uniformed force who may enter any or part of a theatre detail badge worn by such officers or members, if any.

7th. Theatre details shall also, during each tour of duty, inspect the framed instructions with sample theatre detail badge attached (two of which will be furnished to each theatre, with the request to conspicuously display one near the ticket taker's position and the other at the stage entrance), and report whenever they may be found out of place or order.

8th. In signing the report of duty performed at theatres, officers and members shall add the number of the badge worn by them.

9th. The penalty for losing a theatre detail badge is fixed at five dollars.

VIII. — A badge of office of the accompanying design was adopted for the ununiformed force and subordinates of the Department, who may, in the judgement of the Board, require it, the official title to be engraved upon the upper scroll with a number in the number plate where there is more than one of the same grade. The badge is to be used only when necessary as an emblem of their

official authority and position, and not to admit within the lines at fires. They will be issued and a record thereof kept by the Property Record Clerk. The penalty for the loss of the badge is fixed at five dollars.

By order of the Board.
CORNELIUS VAN COTT,
President.

CARL JUSSEN,
Secretary.

Headquarters Department of Fire,

365 and 367 JAY STREET,

BROOKLYN, N. Y., OCTOBER 18th, 1894.

GENERAL ORDERS No. 36,
EIGHTH SERIES.

On and after this date, the insignia of rank of the officers of the department will be metal, gilded, and shall be worn on the cap and coat collar.

By order of

F. W. WURSTER,

Commissioner.

EXCERPT FROM

Charter of the City of New York

§733. Uniforms and badges.

It shall be the duty of the fire commissioner to make suitable regulations under which the officers and men of the fire department shall be required to wear an appropriate uniform and badge, by which in case of fire and at other times, the authority and relations of such officers and men in said department may be known as the exigency of their duties may require. The fire commissioner shall select the grade of cloth and quality required for said uniforms, but shall not prescribe where or of whom said uniforms or uniform clothing shall be purchased, or the price to be paid therefor, and no contractor or agent or employee of any contractor for the making of uniforms for the fire department shall have an office within any building belonging to or under the control of said fire department. It shall be a misdemeanor, punishable by imprisonment for a period of not less than sixty days, for a person not enrolled or employed, or appointed by the said department, to wear the whole or any part of the uniform or insignia prescribed to be worn by the rules and regulations of the fire department or do any act as firemen not duly authorized by the commissioner, or to interfere with the property or apparatus of the fire department in any manner unless by the authority of the fire commissioner. Any person who shall falsely represent any member of the uniformed force of the fire department, or who shall maliciously, with intent to deceive, use or imitate any of the signs, fire caps, badges, signals or devices adopted or used by the said department, shall b deemed guilty of a misdemeanor, and shall be subject to a fine of not less than twenty-five dollars or more than two hundred and fifty dollars, and to imprisonment for a term not less than ten days, or more than three months, such fines when collected, to be paid over to the trustees of the New York fire department relief fund.

Fire Department City of New York.

GENERAL ORDER

On March 15, 1900, by order of the Fire Commissioner, a change will be made in the fire-line badge now in use, and a new one issued by the New York Fire Department for the Boroughs of Manhattan, The Bronx, Brooklyn, and Queens, of the following design:

Any person, except as hereinafter mentioned, not provided with a badge of the above design, on and after that date must be excluded from and not allowed to pass within the police lines at fires.

The fire-line badge does not, under any circumstances, authorize admission to buildings. Any one attempting to use it for that or any other purpose than obtaining admission to streets within the police lines at fires must have the badge taken from him and he ejected from the lines.

This order does not apply to the Mayor, the members of the Municipal Council, heads of Municipal Departments, uniformed firemen and insurance patrolmen, or officials of the Fire Department wearing a badge of the following design:

You will also, on and after March 15th, 1900, take any and all badges formerly issued by the Fire Department of the City of New York, or by the late Fire Department of the City of Brooklyn or Long Island City, authorizing admission within the fire lines, from all persons attempting to make use of or displaying them, and forward such badges to the Central Office for transmission to the Fire Department.

No badge other than those above described are entitled to allow the bearer admission within police lines at fires.

EDWARD F. CROKER,
Chief of Department.

POLICE DEPARTMENT

OF THE CITY OF NEW YORK

300 MULBERRY STREET.

New York, December 8, 1906.

GENERAL ORDER NO. 97.

On and after January 1, 1907 a new Fire Line Badge, issued by the Fire Commissioner, will be in use. This badge will be made of metal with the holder's name printed on it. The badge will be of the same size and coloring as here represented.

No persons not holding one of these badges will be allowed inside the fire lines after that date, except members of the Police and Fire Departments, Post Office Department and Fire Patrol, while in the proper performance of their duty in the vicinity of the fire.

Police cards will not admit holders inside the fire lines.

THEO. A. BINGHAM,
Police Commissioner.

POLICE DEPARTMENT
CITY OF NEW YORK

OFFICE OF THE CHIEF INSPECTOR

New York, September 24, 1918

COMMANDING OFFICER,
Sir:

The above is a photographic reproduction of the shield carried by members of the New York Fire Department Auxiliary Corps. These men are co-operating with the Fire Department and will be permitted to enter within police lines at fires.

This circular to be posted on bulletin board for the information and guidance of the Force.

JOHN DALY,
Chief Inspector

EXCERPT

FIRE DEPARTMENT
CITY OF NEW YORK

RULES AND REGULATIONS
FOR THE GOVERNMENT OF THE OFFICERS AND MEMBERS OF THE
AUXILIARY CORPS
VOLUNTEER SERVICE

Rules and Regulations Auxiliary Corps, F.D., N.Y.

BADGES

Section 32. No member shall dispose of such badge and the loaning of same to any person shall be cause for dismissal.

Should any badge be lost, the loss shall at once be reported to the Sergeants and Commanding Officers of the Companies.

When ceasing to become members of the Auxiliary Corps, badges together with all equipment that may be in possession of the member shall be returned to the Commanding Officer of the Company.

When reporting in at fires members shall display badges and hold themselves in readiness to perform any duty that may be assigned to them.

Efforts must be made at fires to locate the Auxiliary Officer in command so that orders may be transmitted from the Commanding Officer at the fire through him to the Auxiliary members.

Should a member change his address, this fact should be reported to the Sergeant of his company and to the Commanding Officer.

No more than 4 members of the Auxiliary Company shall ride on any apparatus on going to or returning from a fire, except as otherwise ordered.

The term Commanding Officer as used in these rules shall mean the Commanding Officer of the regular Fire Department either in company quarters of at a fire.

Section 33. These rules and regulations shall take effect immediately.

BY ORDER OF
ELI JOSEPH,
Special Deputy Fire Commissioner

Approved by
THOMAS J. DRENNAN, *Fire Commissioner*
JOHN KENLON, *Chief of Department*

EXCERPT FROM

Rules & Regulations of the Uniformed Force of the Fire Department of the City of New York September 1, 1926

Uniform Cap

Sec. 307. Engineers of Steamer - Same as Captains and Lieutenants, excepting trimming of any kind on band, omitting the insignia of rank and substituting therefore a white metal badge 2¾ inches long and 2 inches high, with steam fire engine in an oval, 1¼ inches high and 2⅛ inches long in the centre, surrounded with an ornamental border with the word "Engineer" at the top and the registered number in raised figures below, to be fastened to the cap with an eye under each end, and the lower edge resting upon the upper edge of band.

Firemen of all grades - The same as Engineers of Steamer, omitting the insignia of rank and substituting a white metal Maltese cross, 2 inches square, with the letters "F.D." on the top arm and the letters "N.Y." on the lower arm; a hook and ladder crossed on the right arm and a hydrant on the left arm, with the registered number of each member in raised figures, ⅜ of an inch long, in a circle 1 inch in diameter in the centre, to be fastened to the cap with an eye under each side arm, the lower arm resting upon the upper edge of band.

Insignia of Rank

Sec. 316 Chief of Department - Five trumpets, measuring one and five-sixteenth inches each, crossed, bells outward and projecting beyond mouthpieces so as to form a design one and one-half inches in diameter, to be made of metal and affixed, and to be the property of the Department.

Deputy Chiefs of Department - Same as Chief of Department except that there shall be three trumpets, measuring one and one-half inches each, crossed, with bells downward.

Chiefs of Battalion - Same as Deputy Chiefs of Department except that there shall be two trumpets, crossed, with bells downward.

Captains of Engine Companies (except Captains attached to fireboats) - Same as Chief of Battalion except that the trumpets are to be parallel, with the bells downward, and close together. The insignia for the cap is to have the number of the company in raised letters midway between the trumpets, which are to be placed three-eighths of an inch between bells.

Captains of Rescue Companies - Same as for Captains of engine companies, except that smoke helmet be located in the centre between two life guns, barrels pointing upwards, with roof rope and reel above helmet, life line and slug located below, with number of company on the breathing bag of the helmet.

Captains of Companies Attached to Fireboats - Two parallel trumpets, with bells downward, anchor between trumpets, company number above; to be in silver bullion, on dark blue cloth, surrounded with a bullion cord, size about 1½ by 1⅞ inches.

Captains of Hook and Ladder Companies - Two upright axes, with blades facing each other. The insignia for the cap to have the number of the company in raised figures midway between the axe helves, which are to be placed three-eighths of an inch between helves.

Lieutenants of Engine Companies (except Lieutenants attached to fireboats) - Same as Captains, except that there shall be only one trumpet. The insignia for the caps to have the trumpet placed horizontally, the bell to the left and the number of the company above the centre.

Lieutenants of Rescue Companies - Same as for Captains of Rescue Companies, except that there shall be only one life gun, placed horizontally to the rear of headgear of smoke helmet, barrel pointing to the right.

Lieutenants of Companies Attached to Fireboats - One trumpet horizontally enclosed with a circle 1½ inches in diameter, bell to the left and anchor under trumpet; company's number over trumpet in silver bullion on dark blue cloth, surrounded with a bullion cord. The device to be worn on summer caps shall read as above except "white duck" instead of "blue cloth."

Lieutenants of Hook and Ladder Companies - Same metal as Captain's devices except that there shall be only one axe. The insignia for the cap and collars shall have one axe placed horizontally, the blade to the right. On cap insignia the number of company shall be placed above axe.

Coat Badge

Sec. 318 Chief Officers - A white metal badge, gold plated, with designation of rank and registered number.

Officers - A white metal badge, with designation of rank and registered number.

Engineers of Steamer - Same as uniform cap to be fastened with a pin and improved catch.

Firemen - Same as uniform cap, to be fastened with a pin and improved catch. The coat badge shall be worn on the left breast at the level of the second button from the top, midway between the centre of the body and the side when worn on the coat, and an inch to the right of the left row of buttons on the overcoat.

Ununiformed Firemen - The same as for Uniformed Fireman's badge, except that the words "Ununiformed Fireman" appear in the centre.

Theatre Detail Badge - Members detailed to theatres and places of amusement. A white metal oval, 2⅝ inches long and 2 inches high in the centre, surrounded with an ornamental border, with the words "Theatre Detail" in blue enameled letters on raised ribbon on top, the registered number on small raised oval in centre, and the words, "Fire Department New York" in raised letters below; to be fastened with a pin and improved catch.

Excerpt from 1997 FDNY Uniform Specifications

39

SPECIFICATIONS FOR BADGES

CHIEF OFFICERS AND CHIEF FIRE MARSHAL, DEPUTY CHIEF FIRE MARSHAL, SUPERVISING FIRE MARSHAL, AND FIRE MARSHAL

To be of 16 B.& S. gauge composition metal, gold plated, maltese cross in shape, having four sun ray arms extending from a circle approximately 1⅝ inches in diameter, and having an eagle attached to, and extending above, the upper arm. The seal of the City of New York, in a circle, approximately 1 inch in diameter, shall be mounted in the center and surrounded by a depressed ribbon, approximately ¼ inch wide, with raised beaded borders. The designation of rank and the words "Fire Department New York" shall be mounted on the ribbon and encased with hard fired red French enamel. The registered number, in red enamel, shall be depressed in a raised oval in the center of the upper arm. The badge shall be provided with a pin and an approved catch. All attachments to be hard silver soldered.

COMPANY OFFICERS, PILOTS AND MARINE ENGINEERS

To be of 18 B. & S. gauge composition metal, chromium or rhodium plated over a coat of heavy nickel plating, maltese cross in shape, having four sun ray arms extending from a circle approximately 1⅝ inches in diameter, and having an eagle attached to, and extending above, the upper arm. The seal of the City of New York, in a circle, approximately 1 inch in diameter, shall be mounted in the center and surrounded by a depressed ribbon, approximately ¼ inch wide, with raised beaded borders. The designation of rank and the words "Fire Department New York" shall be mounted on the ribbon and encased with hard fired red French enamel. The registered number, in red enamel, shall be depressed in a raised oval in the center of the upper arm. The badge shall be provided with a pin and an approved catch. All attachments to be hard silver soldered.

FIREMEN

To be of 18 B. & S. gauge composition metal, chromium or rhodium plated over a coat of heavy nickel plating, maltese cross in shape, approximately 2 inches square with the letters "F.D." on the upper arm and the letters "N.Y." on the lower arm; a hook and ladder on the right arm and a fire hydrant on the left arm; with the registered number in raised, separately stamped figures approximately ⅜ inch of an inch high attached by hard silver soldering in a circle approximately 1 inch in diameter in the center. The coat badge to be provided with a pin and an approved catch, and the cap badge with a screw device in the center and a pin or other device to hold badge from turning; all attachments to be hard silver soldered.

SPECIFICATIONS FOR INSIGNIA OF RANK

GOLD PLATED INSIGNIA

To be of 18 B. & S. gauge composition metal, gold plated, round, approximately 1⅝ inches in diameter, trumpets each measuring $1\frac{5}{16}$ inches in length, crossed, bells outward and projecting beyond the mouthpieces, so as to form a design approximately 1½ inches in diameter. Number of trumpets to be according to rank. To be equipped with a screw device in the center and a pin or other device to hold insignia from turning; all attachments to be hard silver soldered.

Chief of Department - To have five trumpets.

Chief in Charge - To have four trumpets.

Assistant Chiefs of Department, Deputy Assistant Chiefs of Department - To have four trumpets.

Deputy Chiefs of Department - To have three trumpets.

Chaplains - To have three trumpets, with the symbol of their religious faith in center across the trumpets.

Battalion Chiefs - To have two trumpets.

40

INSIGNIA FOR CAPTAINS - GENERAL

To be of 18 B. & S. gauge composition metal, chromium or rhodium plated, oval in shape, approximately 2 inches long and 1⅝ inches high; to be equipped with a screw device in the center and a pin or other device to hold insignia from turning; all attachments to be hard silver soldered; design to conform to the following:

Captains of Engine Companies - To have two parallel trumpets, each measuring approximately $1\frac{5}{16}$ inches in length, with bells pointing downward and close together. The cap insignia shall have the number of the company in raised figures, midway between the trumpets, such figures to be separately stamped and attached by hard silver soldering.

Captains of Marine Companies - To have two parallel trumpets, each measuring approximately $1\frac{5}{16}$ inches in length, with bells pointing downward and close together, and the figure of a fouled anchor in an upright position between the trumpets. The cap insignia shall have the number of the company in raised figures, midway between the trumpets above the fouled anchor, such number to be separately stamped figures attached by hard silver soldering.

Captains of Ladder Companies - To have two parallel axes, each measuring $1\frac{5}{16}$ inches in length, with axe pointing upward and blades facing each other. The cap insignia shall have the number of the company in raised figures, midway between the axe helves, such number to be separately stamped figures attached by hard silver soldering.

Captains of Rescue Companies - To have two parallel life guns, each measuring $1\frac{5}{16}$ inches in length, barrels pointing upward, a smoke helmet and breathing bag in the center between the two life guns and a life rope and slug above the helmet. The cap insignia shall have the number of the company in raised figures on the breathing bag, such number to be in separately stamped figures attached by hard silver soldering.

Captains of Squad Companies - To be the same as insignia for Captain of Engine Company, except that the cap insignia shall have the number of the Squad Company preceded by the letter "S" in raised figures, midway between the trumpets, such figures to be separately stamped and attached by hard silver soldering.

INSIGNIA FOR LIEUTENANTS - GENERAL

To be of 18 B. & S. gauge composition metal, chromium or rhodium plated, oval in shape, 1¾ inches long and $1\frac{5}{16}$ inches in height; to be equipped with a screw device in the center and a pin or other device to hold insignia from turning. Company number on cap insignia to be separately stamped figures; all attachments to be hard silver soldered: design to conform to the following:

Lieutenants of Engine Companies - To have one horizontal trumpet, $1\frac{5}{16}$ inches in length, bell pointing outward and carriers upward; the cap insignia shall have the bell of the trumpet to the left of wearer and the number of the company in raised figures in the center, above the trumpet.

Lieutenants of Squad Companies - To be the same as insignia for Lieutenants of Engine Companies, except the cap insignia shall have the number of the Squad Company preceded by the letter "S".

Lieutenants of Marine Companies - To have one horizontal trumpet $1\frac{5}{16}$ inches in length, bell pointing outward and carriers upward; and the figure of a fouled anchor in a horizontal position below the trumpet; the cap insignia shall have the bell of the trumpet to the left of wearer and the number of the company in raised figures, in the center, above the trumpet.

Lieutenants of Ladder Companies - To have one horizontal axe, $1\frac{5}{16}$ inches in length, with blade pointing outward and downward: the cap insignia shall have the blade of the axe to the right of wearer, and the number of the company, in raised figures, in the center, above the axe.

Lieutenants of Rescue Companies - To have one horizontal life gun, $1\frac{5}{16}$ inches in length, with barrel of gun pointing outward, a smoke helmet and breathing bag in the center across the life gun, and a life line and slug above the helmet; the cap insignia shall have the company number, in raised figures, on the breathing bag.

41

INSIGNIA FOR PILOTS, MARINE ENGINEERS

To be of 18 B. & S. gauge composition metal, chromium or rhodium plated, oval in shape, 2 inches long and 1⅝ inches high, having the figure of a fouled anchor $1\frac{7}{16}$ inches long raised and inclined slightly from the horizontal with the flukes facing the left side of the device. For Pilot, a five spoke steering wheel shall be situated upon the center of the device and raised upon the shaft of the anchor. For marine engineer, a three blade propeller shall be situated upon the center of the device and raised upon the shaft of the anchor. The letters "F.D.N.Y." with each letter followed by a period shall be in raised letters approximately ¼ inch of an inch high by $\frac{7}{16}$ inch of an inch long located at the bottom of the device whereon the bade number of the member shall be stamped. Insignia shall be provided with a screw device in the center and a pin or other device to hold insignia from turning; all attachments to be hard silver soldered.

WIPERS

To be of 18 B. & S. gauge composition metal, chromium or rhodium plated over a coat of heavy nickel plating, maltese cross, approximately 2 inches square with the letters "F.D." on the upper arm and the letters "N.Y." on the lower arm; a hook and ladder on the right arm and a fire hydrant on the left arm; with the registered number preceded by the letter "W", in raised, separately stamped figures approximately ⅜ inch of an inch high attached by hard silver soldering in a circle approximately 1 inch in diameter in the center. The coat badge to be provided with a pin and an approved catch, and the cap badge with a screw device in the center and a pin or other device to hold badge from turning; all attachments to be hard silver soldered.

SPECIFICATIONS FOR SHIRT COLLAR INSIGNIA

SHIRT COLLAR INSIGNIA

To be of composition metal, gold, chromium or rhodium plated, design and finish as detailed below; to be equipped with screw device and pin or other device to hold insignia from turning.

Chief of Department - Gold Stars, five stars in a circle, each star approximately $\frac{7}{16}$ inch in diameter.

Chief in Charge - Gold Stars, four stars mounted horizontally, each star approximately ½ inch in diameter.

Assistant Chiefs of Department - Gold Stars, three mounted horizontally, each star approximately ½ inch in diameter.

Deputy Assistant Chiefs of Department - Gold Stars, two approximately ½ inch in diameter.

Deputy Chiefs of Department - Gold Eagle, approximately 1 inch wide by ½ inch high.

Division Commander - Gold Star, one, approximately ½ inch in diameter.

Battalion Chiefs - Gold Oak Leaf, approximately ⅞ inch high by ¾ wide.

Captains - Silver Bars, two, each bar approximately ¾ inch high by ¼ wide, beaded edges, textured finish with 5 diagonal solid stripes across bar, bars to be fixed vertically approximately ¼ inch apart.

Lieutenants - Silver Bar, one, approximately ¾ inch high by ¼ wide, beaded edges, textured finish with 5 diagonal solid stripes across bar.

Pilots - Silver Fouled Anchor, approximately ⅞ inch high by ½ inch wide.

Marine Engineers - Silver Three Bladed Propeller, approximately ⅞ inch in diameter.

Wipers and Firemen - Silver Maltese Cross, approximately ⅞ inch square, with raised letters "City of" on the upper arm and raised letters "New York" on the lower arm; a hook and ladder crossed on the right arm and a fire hydrant on the left arm, and raised letters "F.D." approximately $\frac{3}{16}$ inch high in a circle approximately ⅞ inch in diameter, in the center.

NEW YORK POLICE DEPARTMENT

ARTICLE III

AUTOMOBILE IDENTIFICATION PLATES

1. Section 958 of the Penal Law prohibits the use of any seal, device of arms, sign, lettering or insignia of the United States, or of the State of New York or any political subdivision thereof, on any vehicle not owned or used by the United States, or of the State of New York or a political subdivision thereof.

POLICE COMMISSIONER

2. The following plates have been prescribed and authorized for the purpose of identifying the official automobiles used by the Police Commissioner:

FRONT and REAR

Blue enamel plate with gold border. Stars and letters "N Y" in gold. Letters "P D" and number "1" in white enamel with narrow gold border. Stars, letters and number in raised characters.

23

MANUAL OF PROCEDURE

3. The following plate has been prescribed and authorized by the Police Commissioner for the purpose of identifying Police Department motor vehicles:

(BLUE ENAMELED BACKGROUND, WHITE LETTERS)

4. All "P. D." or other plates heretofore issued for the purpose of identifying Police Department motor vehicles are recalled and all such plates will be forwarded to the Department storehouse.

5. Plates will ordinarily be carried inside of Department cars and will not be displayed on the outside of cars except in cases of emergency when it is necessary to identify them as Police Department cars on official business.

6. "Police" plates are not intended as authority for Police Department cars to fail to comply with traffic regulations, except in cases of urgent police business.

7. Members of the Force observing any plate other than the one described above, or any plate purporting to represent the Police Department of the City of New York but not issued by the Police Department, shall take up and return such plates to the Chief Inspector, together with a report of the name and address of the person from whom taken, and a description and number of the automobile.

8. Commanding Officers will be held strictly accountable for the care and use of automobile identification plates.

FIRE DEPARTMENT AUTOMOBILE IDENTIFICATION PLATES

9. The following sign has been prescribed and authorized by the Fire Commissioner for the purpose of identifying motor vehicles of the Fire Department and the vehicles of the honorary chiefs of that department:

10. Honorary Chiefs of the Fire Department will be permitted to use above sign on other than Fire Department motor vehicles only when in possession of the following permission card, bearing the signature of the Fire Commissioner and the number of such automobile sign:

24

NEW YORK POLICE DEPARTMENT

CARD No.______ SIGN No.______

FIRE DEPARTMENT
CITY OF NEW YORK

I HAVE ISSUED PERMISSION TO

____

THE BEARER OF THIS CARD, TO USE A FIRE DEPARTMENT INSIGNIA ON HIS CAR FOR DEPARTMENTAL PURPOSES ONLY

____
FIRE COMMISSIONER

DATE____________

NOTE—THIS PRIVILEGE IN NO WAY CONFERS UPON THE BEARER THE RIGHT TO VIOLATE THE TRAFFIC REGULATIONS

11. Members of the Police Force observing any sign other than the one described above containing the letters "F. D. N. Y." or any sign purporting to represent the Fire Department of the City of New York but not issued by the Fire Department, shall take up and forward such sign to the office of the Chief Inspector, together with a report of the name and address of the person from whom taken and a description and number of the automobile.

DOCTORS' CARDS

12. A duly licensed physician holding a membership card in The Medical Society of the County of New York will be permitted to leave his automobile on the street in front of premises wherein actually attending a patient, for such length of time as may be necessary for that purpose.

13. The face of such card will certify that the doctor named therein is a member in good standing of The Medical Society of the County of New York, and will bear an imprint of the seal and facsimile of the signature of the Secretary of that Society. The reverse side of card will bear the signature of the doctor to whom the card has been issued.

14. An automobile emblem similar to that of the American Medical Association affixed to the radiator front of the automobile, numbered, displaying the words "Medical Society of the County of New York," or without any lettering, will serve as a further means of identification of the automobile used by such member.

15. Photographic copies of the membership card issued by the Medical Society of the County of New York, and of the automobile emblem, are published herein for the information and guidance of members of the Force:

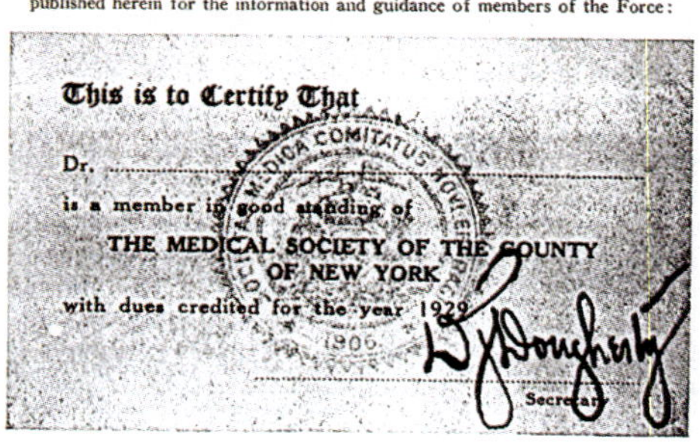

This is to Certify That

Dr. ....

is a member in good standing of

THE MEDICAL SOCIETY OF THE COUNTY OF NEW YORK

with dues credited for the year 1929

Secretary

Facsimile of Front of Card

25

MANUAL OF PROCEDURE

____
Member's Signature

Facsimile of Back of Card

Facsimile of Auto Emblem

16. This order will not be construed so as to permit the use of the streets for the storage of vehicles to which it applies, nor to the leaving of them unattended in front of doctors' offices or residences for long periods of time, when such vehicles are not actually required for use.

17. A member of the Force observing a violation of the restrictions imposed under this order will promptly report the occurrence to his commanding officer, who shall submit report in duplicate of all the facts to the Police Commissioner, through official channels, without delay.

Excerpt from 1929 NYPD Procedures Manual

CHARLES P. GROSS, CHAIRMAN
FRANK X. SULLIVAN, COMMISSIONER
WILLIAM REID, COMMISSIONER

BOARD OF TRANSPORTATION OF THE CITY OF NEW YORK
250 HUDSON STREET
NEW YORK 13, N. Y.
TELEPHONE CANAL 6-6600

WM. JEROME DALY, SECRETARY
HAROLD L. WARNER, COUNSEL
PHILIP E. PFEIFER, GENERAL SUPERINTENDENT

March 15, 1946

FROM: Philip E. Pfeifer, General Superintendent

TO: ALL CONCERNED

SUBJECT: Use of Badges for Transportation on Lines of the New York City Transit System

Recent observations indicate that a number of passengers who ride on lines of the New York City Transit System avoid payment of fare by displaying badges which are not valid for transportation on those lines.

The only badges which are valid for transportation on lines of the New York City Transit System are those issued to members of the New York City Police and Fire Department Forces. The members of those Forces are authorized to enter all stations and passenger vehicles of the New York City Transit System, without payment of cash fare, provided they are in full uniform or provided they properly display their official badge so that it can be seen clearly by the respective fare-collecting Agents of the Board.

Railroad Clerks, Operators and other Agents of the Board of Transportation charged with the responsibility of collecting fares must closely scrutinize all badges presented in payment of fare and the only ones which they are authorized to honor for transportation purposes are the types shown on pages 2 and 3 of this Notice. In cases where a badge not valid for transportation is presented, the Railroad Clerk or Operator should make a record of the number and other descriptive information of such badge and immediately transmit same through regular channels, so that the occurrence may be called to the attention of the proper authorities. Under no circumstances should an employee of the New York City Transit System confiscate a badge, even though the holder thereof improperly used it in an attempt to enter a station or vehicle without payment of fare.

Please give this matter your careful attention and earnestly cooperate with the Board of Transportation in its efforts to stop the fraudulent use, for fare payment purposes, of badges which are not valid for transportation on lines of the New York City Transit System.

Philip E. Pfeifer
General Superintendent.

REF. 501-S

FIRE DEPARTMENT BADGES

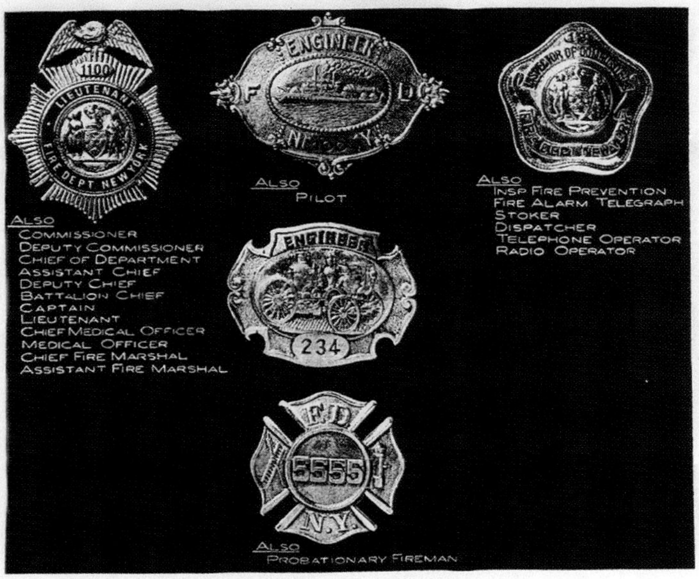

POLICE DEPARTMENT BADGES

RE: OFFICIAL BADGES

For the benefit of those members who have not been in touch with developments about the official recognition of the Club as an "adjunct of the Fire Department," a resume of what has taken place would be in order. For those who do know, this letter will serve to acquaint them with the procedure to be followed to secure a badge.

True to the promise to see what could be done to recognize the Club as a group, which was made to us at our last annual dinner, Commissioner Quayle wrote to your President on July 24 and said that the Mayor had granted his request to recognized the Club as an "adjunct of the Fire Department." This recognition has four reasonable conditions as follows:

1. Preparation of a distinctive badge to be okayed by the Department and when approved, to be numbered. (Approval since received.)

2. An overall ceiling of one hundred badges to be issued and the names and numbers of the holders thereof to be recorded with the Department.

3. Signing of an individual sworn affidavit from each badge holder, releasing the city from any liability in connection with injuries sustained while using the badge.

4. A requirement that any additional members taken into the Club continue to be "representative of outstanding members of our community."

After considerable delay in design and production of the sample due to uncertainty as to what was meant by "distinctive," a very attractive badge has now been approved. It is gold. It has a maltese background with the entire cross fluted. Across the top segment of the cross is a band in which the letters F B C appear. The circular center section occupies about two thirds of the entire diameter of the body (2 1/4".) The center consists of the city seal one inch in diameter. Completely surrounding the seal is a quarter inch circular band. Across the top half of this band will appear the words "Fire Department." Across the bottom half, "City of New York." A small number plate will be attached to the bottom segment of the cross. All letters will be depressed and colored blue. The background of the city seal will also be blue. The badge will be official and will be recorded with the Police Department. The release is a standard Fire Department form originally drawn up for signature of firemen from other cities taking the course in the Fire College and responding with the apparatus. These have to be signed in duplicate - one to be sent to the Chief of Department.

Complete responsibility for determining who, within the membership of the Club, shall get or retain the badge rests with the Board of Directors. The Directors have approved regulations and a form of agreement to be signed by the individual member. This agreement provides that possession of a badge is a privilege of members in the Club and not a right, that each member taking a badge agrees to sign the release approved by the city, that the cost of the badge will be borne by the member, but the badge will be the absolute property of the Club and the cost is not refundable. Other reasonable regulations regarding its use, loss or damage or replacement are included in this agreement. Provision for its surrender on demand of the proper authorities, on leaving the Club or death of a member, are also included.

Numbers will be assigned by lot among the original members and by order of entry thereafter.

Members will have the choice of either a gold plated badge at \$6 or a rolled gold badge at \$18, 20% tax included in each case, including the carrying case. An order blank is enclosed. Please indicate your wishes even if you do not want a badge. All orders must be in within fifteen days after the date of this letter. Do not place the order unless you are certain that you will have no objection to either the city affidavit or the club regulations. A copy of each may be examined at the Club. For the benefit of those members whose address is out of town, a copy of each is enclosed. Payment must be made at time of delivery and delivery will not be completed unless all agreements are signed. Delivery has been promised to us from four to six weeks after the order has been placed with the manufacturer. Members will be notified when their badge is ready.

Please complete the order promptly so that a consolidated order may be placed as soon as possible.

Fraternally

Joseph F. Hirschmann
President

New York, N. Y. September 25, 1946

Letter to Fire Bell Club members regarding the issuance of badges.

The City of New York

FIRE DEPARTMENT
250 LIVINGSTON STREET BROOKLYN, N.Y. 11201-5884

BUREAU OF PERSONNEL

TO: Thomas J. VonEssen Fire Commissioner

FROM: ____________

SUBJECT: Retention of official F.D.N.Y. badge upon retirement after at least 30 years of service.

NAME: ____________

RANK: ____________

UNIT: ____________

BADGE #: ____________

DATE APPT.: ____________

DATE RETIRE: ____________

SS# ____________

REMARKS:

Respectfully Submitted

____

COMPANY COMMANDER: ____________
Unit Date

BATTALION CHIEF: ____________
Unit Date

DEPUTY CHIEF: ____________
Unit Date

Letter requesting retention of badge for retiring member with 30 years of service.

Names of Volunteer Fire Department Companies in Greater New York City

During the volunteer era, it was common for companies to take on a name. The following is a list of company names from the first days of the Department to the present. The dates indicate the organization and disbandment dates of each company and, in some cases the dates upon which names were changed. This will help determine the dates of any badges that display a company name. The name in italics is the company nickname.

MANHATTAN

Engine Companies

No.	Name	Dates
1 -	Hudson (*Hayseed*)	00/00/1731 - 08/05/1846
		12/26/1846 - 09/08/1865
2 -	Chatham	00/00/1731 - 09/22/1845
	Excelsior (*Quaker*)	01/06/1846 - 09/08/1865
3 -	New York	<1783 - 00/00/1812
	Dugan	
	Metamora	03/22/1830 - 12/06/1843
	Hope	07/16/1847 - 00/00/1852
	National/Broderick	10/25/1852 - 09/22/1845
	Forrest	03/05/1860 - 09/08/1865
4 -	American	00/00/1749 - 09/22/1845
	Niagara	03/05/1849 - 09/08/1865
5 -	Protection (*Honeybee*)	00/00/1762 - 09/08/1865
6 -	Neptune (*Bean Soup*)	00/00/1765 - 08/05/1846
	Americus (*Big 6, Tiger*)	12/11/1849 - 09/08/1865
7 -	Lafayette	00/00/1701 - 03/01/1843
	Hamilton	10/04/1843 - 02/04/1847
	Lexington	12/26/1849 - 09/08/1865
8 -	Manhattan (*Elephant*)	00/00/1776 - 04/04/1831
	Essex	04/04/1831 - 09/08/1865
9 -	Bolivar	00/00/1783 - 03/01/1843
	Fifteenth Ward	10/27/1837 - 06/22/1843
	Dolphin	02/01/1843 - 12/26/1846
	United States (*Rock*)	08/14/1849 - 12/00/1854
	Marion	12/00/1854 - 09/08/1865
10 -	Niagara	00/00/1783 - 02/04/1836
	Water Witch	01/04/1837 - 03/01/1843
		03/24/1843 - 08/01/1855
	United Water Witch	09/28/1855 - 09/08/1865
11 -	Oceanus	12/28/1780 - 09/08/1865
12 -	Knickerbocker (*Old Nick*)	<1783 - 03/01/1843
	Tradesman's	02/22/1844 - 04/29/1847
	Knickerbocker	01/28/1847 - 09/08/1865
13 -	Eagle	<1783 - 09/08/1865
14 -	Columbian	<1783 - 09/08/1865
15 -	Peterson (*Old Maid*)	00/00/1785 - 09/27/1849
	Wreath of Roses	03/03/1852 - 07/11/1855
	Chatham/Hibernia	10/09/1855 - 09/08/1865
16 -	Aetna (*Hounds*)	00/00/1803 - 00/00/1833
	Chelsea	02/05/1834 - 05/04/1836
	Croton (*Cit*)	05/04/1836 - 05/06/1841
		05/10/1842 - 09/18/1852
	Gotham	09/24/1854 - 00/00/1857
	Mohawk	00/00/1857 - 09/08/1865
17 -	Scott Mutual	12/00/1787 - 06/22/1842
	East River	02/13/1852 - 09/08/1865
18 -	Union (*Shad Belly*)	04/30/1782 - 00/00/1828
	Union (*Dry Bones*)	08/06/1829 - 00/00/1846
	Atlantic	03/02/1852 - 12/00/1854
	Union	09/03/1856 - 09/08/1865
19 -	Lafayette	06/25/1792 - 09/08/1865
20 -	Washington	06/25/1792 - 09/08/1865
21 -	Fulton	06/00/1795 - 09/08/1865
22 -	Phenix	00/00/1796 - 00/00/1839
	Protector (*Old Junk*)	06/25/1840 - 09/08/1865
23 -	United States	00/00/1797 - 00/00/1849
	Waverly	03/02/1852 - 09/08/1865
24 -	Jackson	00/00/1798 - 09/08/1865
25 -	Merchants'	00/00/1801 - 09/24/1836
	Cataract	01/30/1839 - 09/08/1865
26 -	Jefferson (*Blue Boys*)	00/00/1803 - 09/22/1845
		12/02/1851 - 09/08/1865
27 -	North River	00/00/1803 - 03/01/1843
	Fort Washington	11/18/1852 - 09/08/1865
28 -	Mechanics'	00/00/1805 - 01/12/1841
	Pacific	01/27/1852 - 09/08/1865
29 -	Constitution (*Rooster*)	00/00/1803 - ?
	Trident (*Old Jeff*)	? - 00/00/1836
	Guardian (*Iron Boys*)	00/00/1836 - 09/08/1865
30 -	Tompkins	00/00/1804 - 05/09/1854
	North River	07/15/1858 - 00/00/1859
31 -	Hope	00/00/1805 - 09/11/1854
	Adriatic	02/24/1857 - 00/00/1858
	Peterson	00/00/1858 - 09/08/1865
32 -	Bunker Hill	00/00/1807 - 05/06/1858
	Pete Masterson	12/20/1859 - 09/08/1865
33 -	Bombazula	00/00/1807 - 00/00/1828
	Black Joke	00/00/1828 - 11/06/1844
		03/06/1852 - 09/08/1865
34 -	Howard (*Red Rover*)	00/00/1807 - 09/08/1865
35 -	Columbus	00/00/1807 - 09/08/1865
36 -	Equitable	08/13/1810 - 08/05/1846
	Harry Howard	03/10/1848 - 09/08/1865
37 -	Tradesman's	00/00/1811 - 03/01/1843
		09/26/1853 - 09/08/1865
38 -	No name	00/00/1811 - 07/09/1838
	Southwark	02/10/1840 - 09/08/1865
39 -	Franklin	04/06/1812 - ?
	Good Intent (*Skiver*)	03/20/1820 - 03/01/1843
	Franklin	12/27/1853 - 09/08/1865
40 -	Lady Washington (*White Ghost*)	00/00/1812 - 12/06/1843
	(*White Gal*)	12/19/1853 - 09/08/1865
41 -	Clinton (*Old Stag*)	03/08/1813 - 09/08/1865
42 -	Floating Engine	00/00/1800 - 00/00/1824
	East River	06/21/1824 - 00/09/1838
	Empire (*Haywagon/Mankiller*) /Northern Liberty	03/05/1840 - 09/08/1865
43 -	Manhatta	10/29/1821 - 09/08/1865
44 -	Live Oak (*Old Turk*)	08/02/1824 - 09/08/1865
45 -	Yorkville	09/11/1826 - 00/00/1845
	Aurora	00/00/1845 - 09/08/1865
46 -	Relief (*Bull's Head*)	10/08/1827 - 10/07/1854
	Valley Forge	04/21/1860 - 09/08/1865
47 -	Mechanics'	03/10/1828 - 06/22/1842
	Aqueduct	03/06/1854 - 09/18/1855
	New York	06/04/1860 - 03/23/1864
	Croton	05/26/1864 - 09/08/1865
48 -	Mazeppa	11/17/1828 - 09/08/1865
49 -	No name	04/04/1832 - 00/00/1835
	Pocahontas	00/00/1835 - 09/08/1865
50 -	Mohegan	04/01/1840 - 03/10/1848
	Lone Star	09/01/1854 - 00/00/1858
	Liberty	00/00/1858 - 02/11/1864
51 -	Mutual	12/22/1854 - 09/08/1865
52 -	Undine	02/19/1863 - 09/08/1865
53 -	Hudson River	02/19/1863 - 12/17/1865
		06/13/1865 - 09/08/1865
54 -	Paulding	02/19/1863 - 02/11/1864
— -	Supply Engine	00/00/1827 - 02/26/1841
— -	Exempt Engine Company	12/27/1854 - 09/08/1865

Hose Companies

No.	Name	Dates
1 -	Eagle	09/07/1812 - 10/00/1865
2 -	Niagara/Knickerbocker	12/05/1831 - 09/08/1865
3 -	Independence/Independent	11/19/1832 - 09/08/1865
4 -	Lafayette	02/05-1833 - 00/00/1850
	Marion (*Veto*)	00/00/1850 - 09/08/1865
5 -	New York	03/20-1833 - 09/23/1836
		09/26/1836 - 06/04/1860
	Edwin Forrest	12/19/1860 - 09/08/1865
6 -	Croton	07/06/1837 - 09/08/1865

No.	Name	Dates
7 -	Wave	12/05/1836 - 04/02/1845
	Ringgold	09/22/1845 - 09/08/1865
8 -	First Ward	10/15/1836 - 10/15/1852
	City	10/15/1852 - 09/08/1865
9 -	Columbian	01/17/1837 - 09/08/1865
10 -	Liberty/East River	02/01/1837 - 09/08/1865
11 -	Gulick	03/25/1837 - 09/08/1865
12 -	Washington	03/24/1837 - 01/21/1858
	Minute	01/20/1859 - ?
	Washington	? - 09/08/1865
13 -	Express	05/06/1837 - 12/06/1843
	Jackson	02/07/1844 - 09/08/1865
14 -	Atlantic (*Lady Suffolk*)	05/03/1837 - 09/09/1857
	Excelsior	03/11/1858 - 09/08/1865
15 -	Victory	05/23/1837 - 00/00/1852
	Fulton	00/00/1852 - 00/00/1856
	Peterson	00/00/1856 - 04/06/1859
	Atlantic	07/04/1859 - 09/08/1865
16 -	Naiad	11/21/1837 - 01/10/1848
	Tompkins	05/23/1848 - 09/08/1865
17 -	Clinton	01/27/1838 - 09/08/1865
18 -	Franklin	04/11/1838 - 09/08/1865
19 -	American	01/29/1839 - 09/08/1865
20 -	Second Ward	10/01/1836 - 06/22/1842
	Humane	06/22/1842 - 09/08/1865
21 -	Protection	02/13/1839 - 03/20/1846
	Hudson	10/21/1846 - 09/08/1865
22 -	Phenix	08/01/1839 - 09/08/1865
23 -	Triton	04/01/1839 - ?
	Perry	? - 09/08/1865
24 -	Eighth Ward	10/03/1836 - 06/22/1842
	National	06/22/1842 - 09/08/1865
25 -	United States (*Fashion*)	05/27/1939 - 00/00/1841
		07/13/1843 - 09/08/1865
26 -	Rutgers/Van Buren	06/01/1839 - 09/08/1865
27 -	Third Ward	02/00/1836 - 10/21/1846
	Neptune	11/16/1847 - 09/08/1865
28 -	Fifth District	05/25/1839 - 06/22/1844
	Pearl	06/22/1844 - 09/08/1865
29 -	Eleventh Ward/Mayflower/Union	06/22/1842 - 09/18/1854
	Metamora/Continental	12/27/1854 - 00/00/1863
30 -	No name	06/01/1840 - 02/05/1843
	Edwin Forrest	07/06/1847 - 00/00/1850
	Laurel	00/00/1850 - 09/08/1865
	George B. McClellan	00/00/1863 - 09/08/1865
31 -	Putnam	04/28/1842 - 09/08/1865
32 -	Richard M. Johnson	06/01/1840 - 12/02/1851
	Index	00/00/1852 - 09/08/1865
33 -	Hoffmire	00/00/1839 - 02/02/1841
	City (*Kentuck*)	02/03/1841 - 00/00/1849
	Warren (*Black Bud*)	00/00/1849 - 09/08/1865
34 -	Star	06/22/1842 - ?
	Lafayette	? - 05/05/1860
	Hibernia	05/25/1863 - 02/08/1864
35 -	Fifteenth Ward	06/22/1842 - 00/00/1849
	Baltic	00/00/1849 - 06/30/1859
	Crystal	06/25/1863 - 09/08/1865
36 -	Oceana	05/05/1845 - 09/08/1865
37 -	No name	06/22/1842 - 02/29/1843
	Madison	00/00/1848 - 09/08/1865
38 -	Amity	03/08/1842 - 09/08/1865
39 -	Metropolitan	05/08/1842 - ?
	Mohawk	? - 09/08/1865
40 -	Empire (*Red Bird*)	00/00/1843 - 09/08/1865
41 -	Alert	03/24/1843 - 09/08/1865
42 -	No name	06/01/1839 - 06/01/1840
	Mazeppa	10/12/1848 - 05/14/1850
	Mount Pleasant	00/00/1850 - 09/08/1865
43 -	No name	05/20/1839 - 04/28/1842
	Pioneer	09/28/1848 - 09/08/1865
44 -	Eleventh Ward	06/01/1839 - 06/22/1842
	Washington Irving	02/01/1849 - 00/00/1864
45 -	Red Jacket	12/10/1849 - 07/19/1858
	C. Godfrey Gunther	06/25/1863 - 09/08/1865
46 -	Nassau	11/05/1849 - 09/18/1854
	Valley Forge	12/20/1854 - 04/21/1860
47 -	Mechanics' Own/Mechanics	04/01/1850 - 09/08/1865
48 -	Carson	11/00/1850 - 00/00/1851
	Americus	00/00/1851 - 09/08/1865
49 -	Lady Washington	00/00/1851 - 09/08/1865
50 -	Corlies	11/00/1850 - ?
	Hope	09/08/1851 - 09/08/1865
51 -	Relief	09/01/1852 - 09/08/1865
52 -	Undine	10/27/1851 - 02/19/1863
53 -	Naiad	01/21/1852 - 12/01/1858
		09/00/1859 - 09/08/1865
54 -	Eureka	02/19/1852 - 07/15/1858
55 -	Harry Howard	09/17/1853 - 09/08/1865
56 -	Equitable	03/05/1853 - 12/20/1855
	Nassau	11/27/1856 - 09/08/1865
57 -	Paulding	01/27/1854 - 02/19/1863
58 -	Merchant	05/14/1856 - 12/19/1860
59 -	Ion	09/22/1856 - 00/00/1859
	Manhattan	00/00/1859 - 09/08/1865
60 -	M. T. Brennan	03/28/1854 - 09/08/1865
61 -	Zephyr	12/31/1856 - 09/08/1865
62 -	Minute	02/23/1857 - 01/20/1859
— -	Fifth Ward Exempt	02/00/1860 - 09/08/1865

Hook & Ladder Companies

No.	Name	Dates
1 -	Mutual	07/10/1772 - 09/08/1865
2 -	No name	< 1783 - 08/01/1850
	Chelsea	09/13/1851 - 00/00/1864
3 -	Phoenix	09/12/1804 - 09/08/1865
4 -	Eagle	10/21/1811 - 00/00/1864
	George B. McClellan	00/00/1864 - 09/08/1865
5 -	Union (*Screamer*)	00/00/1826 - 09/08/1865
6 -	Lafayette	07/27/1829 - 09/08/1865
7 -	Mechanics	09/07/1837 - 09/08/1865
8 -	Empire	09/06/1848 - 09/08/1865
9 -	America	10/07/1851 - 12/26/1855
	Washington	07/10/1856 - 09/08/1865
10 -	Narragansett	07/01/1839 - 06/00/1859
	C.V. Anderson	06/00/1859 - 09/08/1865
11 -	Knickerbocker	03/23/1852 - 07/12/1855
	Harry Howard	08/26/1857 - 09/08/1865
12 -	Friendship	02/26/1852 - 09/08/1865
13 -	Marion	08/17/1852 - 09/08/1865
14 -	Columbian (*Wide Awake*)	05/11/1854 - 09/08/1865
15 -	Baxter	08/00/1855 - 09/08/1865
16 -	Manhattan	10/10/1859 - 00/00/1860
	Liberty	00/00/1860 - 09/08/1865
17 -	John Decker	03/05/1860 - 09/08/1865
18 -	Hibernian	08/08/1860 - 09/08/1865

Hydrant Companies

No.	Dates
1	00/00/1832 - 04/21/1855
2	06/01/1839 - 04/21/1855
3	06/01/1839 - 12/05/1844
	06/13/1847 - 04/21/1855
4	04/18/1853 - 04/21/1855

BROOKLYN

Township of Breuklin	1785-1823
Village of Brooklyn	1823-1834
City of Brooklyn	1834-1855
Brooklyn F.D. Western District	1855-1869
Brooklyn Fire Department	1869-1898

The following volunteer company names were discontinued in 1869, but many companies remained active in the paid department.

Alert Hose 3	? - 09/15/1869
American Engine 9	09/24/1868 - 09/15/1869
Americus Hose 7	11/27/1835 - 06/01/1898
Atlantic Hose 1	00/00/1835 - 00/00/1869
Bedford Hose 11	? - 10/03/1867
	05/21/1868 - 09/17/1868
	10/05/1868 - 06/01/1869

Brooklyn Engine 17	09/28/1848 - 06/01/1898
Carroll Hose 15	10/15/1863 - 01/30/1865
Cashow Hose 12	? - 01/30/1865
Clinton H&L 2	01/30/1840 - 09/15/1869
Columbia Engine 10	00/00/1839 - 06/01/1869
Constitution Engine 7	10/24/1828 - 09/15/1869
Continental Engine 9*	
Crystal Hose 4*	
Degraw H&L 4	00/00/1861 - 06/01/1869
Dutch Kills Engine 15*	
Eagle Engine 4	00/00/1812 - 02/11/1864
Empire Engine 19*	
Empire H&L 3	10/04/1838 - 06/01/1869
Eureka Hose 14*	
Frontier Hose 5	00/00/1844 - 01/20/1869
Franklin Engine 3	00/00/1817 - 09/15/1869
Good Will Engine 4	04/12/1864 - 06/01/1869
Hibernia Engine 16	10/03/1867 - 06/01/1869
Hope Hose 9	00/00/1861 - 09/24/1868
Jackson Engine 11	? - 09/15/1869
Lafayette Engine 5	06/08/1825 - 09/15/1869
Mechanic Hose 2	? - 09/15/1869
Montauk Engine 22	02/22/1853 - 06/01/1869
Montauk Hose 4	05/10/1853 - 11/06/1855
Montross (later Lafayette) H&L 1	10/00/1817 - 01/01/1898
Mount Prospect Engine 16	? - 06/30/1865
Mount Vernon Hose 10*	
Myrtle Hose 17	04/18/1864 - 05/20/1869
Neptune (*Rooster*) Engine 2	00/00/1787 - 09/15/1869
Niagara Engine 8	09/24/1868 - 09/15/1869
Niagara Hose 11	00/00/1861 - ?
Pacific Engine 14	09/19/1825 - 01/01/1898
Phoenix Engine 12	? - 09/15/1869
Protector (*Old Bean Soup*) Engine 6	09/02/1825 - 09/15/1869
Prospect Heights Engine 1	
Putnam Engine 21	? - 09/15/1869
Rescue H&L 5	? - 09/15/1869
Sprague H&L 6*	
Valley Forge Engine 11	
Victory Engine 13	
Volunteer Engine 8	00/00/1853 - ?
Washington Engine 1	04/30/1785 - 06/01/1889
Washington Hose 6	00/00/1853 - 06/01/1889
Water Witch Hose 8*	
Union Hose 5	
Zephyr (*Swamp Fog*) Hose 4	00/00/1865 - 09/15/1869

City of Williamsburgh F.D.	1823 - 1855
(Prior to 1823 Fire Department history is unknown)	
Brooklyn F.D. Eastern District	1855 - 1869
Brooklyn Fire Department	1869 - 1898

Volunteer company names were discontinued in 1869 but many companies remained active in the paid department.

Atlantic Hose (and Relief) 2	11/27/1835 - 09/15/1869
Continental Engine 9	00/00/1855 - 09/15/1869
Eagle (Bucket) Engine 6	
Excelsior Hose 9	? - 06/30/1865
Friendship Hose 3	00/00/1855 - 09/15/1869
Good Intent (*The Burg*) Engine 3	00/00/1844 - 09/15/1869
Good Will Hose 4	? - 06/01/1889
Lady Washington (*Rooster*) Engine 1	03/00/1834 - ?
Marion Hose 1	? - 06/30/1865
Mutual H&L 1	06/00/1836 - 00/00/1855
Neptune Engine 7	00/00/1855 - 09/15/1869
North Liberties Engine 5	00/00/1836 - 09/15/1869
Pacific Engine 8	? - 06/30/1865
Protection (*White Boy*) Engine 2	00/00/1834 - 09/15/1869
Putnam H&L 2	00/00/1855 - 09/15/1869
Red Jacket Engine 10	
Union Engine 5	
United States Engine 4	
Victory Hose 8	? - 09/15/1869
Young America H&L 3	

Barren Island F.D.	? - 00/00/1930

Bay Ridge F.D.	circa 1887 - 1891
Bay Ridge Engine 1	
Neptune Engine 2	

Coney Island F.D.	
Atlantic Engine 1	03/12/1892 - 12/09/1895
Atlantic Hose 1	12/09/1886 - 12/09/1895
Atlantic Ladder 1	12/09/1886 - 12/09/1895
Hose 2	05/12/1892 - 12/09/1895

East New York F.D.	
American Eagle Engine 1	00/00/1850 - 00/00/1862
(Hope, Americus, Empire, Hausel, Big Six)	
Eagle Ladder 1	? - 00/00/1862
Liberty Hose 1	? - 00/00/1862
Union Ladder 1	00/00/1850 - 00/00/1862

Flatbush F.D.	
Farmers Ladder 2	09/01/1888 - 01/20/1896
Malbone Hose 5	01/26/1889 - 01/20/1896
Melrose Hose 2	02/19/1886 - 01/20/1896
Prospect Ladder 3	02/03/1893 - 01/20/1896
Washington Engine 1	03/31/1821 - 01/20/1896
Washington Ladder 1	00/00/1865 - 01/20/1896
Windsor Terrace Hose 3	01/11/1887 - 01/20/1896
Woodbine Hose 4	02/01/1887 - 01/20/1896

Flatlands F.D.	
Canarsie Hose 1	10/14/1891 - 12/15/1897
Flatlands Engine 1	00/00/1859 - 12/15/1897
Flatlands Ladder 1	05/14/1890 - 12/15/1897

Gerrittsen Beach F.D.	00/00/1922 - Active

Gravesend F.D.	
Gravesend Ladder 1	? - 12/09/1895
Union Engine 1	? - 12/09/1895

New Lots F.D.	
Alert Pump & Bucket 1	00/00/1875 - 08/04/1886
Alert Hose 3	00/00/1875 - 08/04/1886
American Eagle Engine 1	08/15/1850 - 08/04/1886
Eagle Ladder 2	00/00/1862 - 08/04/1886
Empire Engine 1	00/00/1864 - 08/04/1886
Franklin Engine 2	00/00/1872 - 08/04/1886
Hope Hose 2	? - 08/04/1886
Independent Pump & Bucket Co.	00/00/1875 - 08/04/1886
Liberty Hose & Bucket 1	00/00/1862 - 08/04/1886
Neptune Engine 1	00/00/1875 - 08/04/1886
Union Ladder 1	00/00/1862 - 08/04/1886

New Utrecht F.D. (Organized 1842)	
Bay Ridge Engine 1	? - 02/01/1896
Bay Ridge Ladder	? - 02/01/1896
Blythe Bourne Engine 5	? - 02/01/1896
Hamilton Engine 1 (Fort Hamilton)	? - 02/01/1896
Hamilton Ladder 3 (Fort Hamilton)	? - 02/01/1896
Liberty Ladder 1 (Bath Beach)	? - 02/01/1896
Neptune Engine 2 (Bay Ridge)	? - 02/01/1896
Nesho Ladder	? - 02/01/1896
Old Jackson Hose	? - 02/01/1896
Old Jackson Ladder 4	? - 02/01/1896

Plum Beach F.D.	? - 00/00/1930

Sheepshead Bay F.D.	
Friendship Engine 1	? - 12/09/1895
Friendship Hose 1	? - 12/09/1895
Friendship Ladder 1	02/04/1889 - 12/09/1895
Harry Howard Hose 1	08/12/1893 - 12/09/1895

* - Listed in 1867 Annual Report

QUEENS

Astoria F.D.	1839 - 1873
Protection Engine	
Bayside F.D.	
Volunteer Engine 1	? - 11/15/1924
Enterprise Ladder 1	00/00/1898 - 11/15/1924
Black Stump F.D.	
Black Stump Ladder 1	00/00/1905 - 10/16/1923
Broad Channel Volunteer F.D.	00/00/1908 - Active
Brooklyn Hills F.D.	? - 1906
Deraismus Engine & Hose	
Breezy Point F.D.	1910 - Active (Currently Point Breeze F.D.)
Creedmore F.D.	
Creedmore Hose 1	00/00/1907 - 10/16/1923
College Point F.D.	
Eagle Ladder 1	02/17/1863 - 12/01/1908
Enterprise Hose 2	00/00/1863 - 12/01/1908
Touch Down Steamer	? - 12/01/1908
Union Engine 1	06/30/1860 - 03/01/1864
Union Bucket Company 1	02/17/1857 - 06/30/1860
Union Hose 1	02/17/1857 - 12/01/1908
Columbia F.D.	1901 - 1908
Corona F.D.	? - 00/00/1889
Pioneer Ladder (Became part of Newtown F.D. in March 1890)	
Douglaston F.D.	
Douglaston Hose 1	09/23/1903 - 11/30/1929
Elmhurst F.D.	? - 1906
East Hamilton Beach F.D.	1910 - 1930
Far Rockaway (Oceanus) F.D.	
Arverne Engine 2	12/20/1893 - 09/01/1905
Arverne Hose 2	03/24/1887 - 09/01/1905
Atlantic Ladder 1	04/08/1875 - 09/01/1905
Defender Hose	? - 09/01/1905
Mohawk Hose 2	02/09/1888 - 09/01/1905
Oceanic Hose 1	10/13/1886 - 09/01/1905
Protective Engine 1	01/26/1887 - 09/01/1905
Protective Ladder 1	01/26/1887 - 09/01/1905
Floral Park F.D.	
Floral Park Ladder 1	
Flushing F.D.	
Empire Hose 1	06/15/1856 - 12/01/1908
Flushing Hose 3	06/00/1889 - 12/01/1908
Murray Hill Hose 4	04/20/1894 - 12/01/1908
Mutual Engine 1	05/19/1860 - 12/01/1908
Rescue Ladder 1	05/18/1854 - 12/01/1908
Young America Engine 2	? - 12/01/1908
Young America Hose 2	05/18/1854 - 12/01/1908
Forest Hills F.D.	
Forest Hills Fire Company	00/00/1917 - 11/15/1924
Hollis F.D.	
Columbia Hose 1	00/00/1890 - 01/15/1921
Hollis Ladder 1	00/00/1890 - 01/15/1921
Jamaica F.D.	
Atlantic Ladder 1	00/00/1797 - 07/05/1907
Continental Bucket	00/00/17??
Continental Ladder 2	04/05/1860 - 07/05/1907
Degraw Hose 1	00/00/1871 - 00/00/1887
Distler Hose 3	00/00/1889 - 07/05/1907
Eagle Hose 1	00/00/1860 - 00/00/1871
Excelsior Hose 2	00/00/1887 - 07/05/1907
Fosdick Hose 4	00/00/1889 - 07/05/1907
Jamaica Engine 1	00/00/1849 - 00/00/1851
Jamaica Engine 2	1849
Jamaica Hose 6	? - 07/05/1907
Jamaica Ladder 3	00/00/1849 - 07/05/1907
Neptune Engine 2	00/00/1852 - 00/00/1888
Resolute Hose 5	00/00/1891 - 07/05/1907
Woodhull Hose 1	00/00/1888 - 07/05/1907
Jamaica South F.D.	1905 - 1912
Little Neck F.D.	
Active Hook & Ladder 1	00/00/1905 - 11/30/1929
Long Island City Volunteer F.D.	(? - 1891)

Company names were discontinued on disbandment of volunteer department on February 1, 1891.

Active Ladder 1	
Americus Engine 6	
Astoria Engine 1	
Columbia Ladder 4	
Empire Hose 2	
Franklin Engine 3	
Friendship Ladder 3	
Hope Hose 4	
Hunter (Dutch Kills) Engine 4	
Jackson Engine 5	
Jackson Hose 5	
Live Oak Ladder 2	
Mohawk Hose 1	
Protection Engine 2	
Rambler Hose 3	
Steinway Hose 7	
Tiger Hose 8	
Union Hose 6	
Washington Engine 7	
Middle Village F.D.	? - 1889
Morris Park F.D.	? - 1890
Newtown F.D.	
Eagle Ladder 2	? - 02/20/1913
Elmhurst Ladder 11	10/22/1896 - 02/20/1913
Fearless Ladder 7	07/01/1890 - 09/01/1913
Glendale Engine 9	09/10/1896 - 09/01/1913
Glendale Park Ladder 13	09/13/1902 - 09/01/1913
Gooderson Engine 2	05/02/1862 - 02/20/1913
Ivanhoe Hose 10	07/25/1896 - 07/02/1907
Ivanhoe Ladder 10	10/02/1896 - 02/20/1913
Jackson Hose 3	?
Louna Engine 8	06/08/1896 - 02/20/1913
Maspeth Engine 4	06/00/1888 - 02/20/1913
Maspeth Ladder 5	03/15/1889 - 02/20/1913
Metropolitan Engine 12	08/28/1902 - 09/01/1913
Old Friendship Hose 3	? - 1913
Pioneer Hose	? - 02/20/1913
Pioneer Ladder 8	03/00/1890 - 09/01/1913
Wandowenck Engine 1	00/00/1843 - 09/01/1913
Wandowenck Ladder 1	04/15/1845 - 02/20/1913
Woodside Ladder 3	05/00/1878 - 02/20/1913
North Beach F.D.	
Bowery Bay Hose Co.	00/00/1908 - 00/00/1921
Point Pleasant F.D.	? - 01/01/1915
Point Pleasant Ladder 1	
Queens F.D.	
Queens Hose 2	00/00/1902 - 05/21/1921
Queens Ladder 1	04/29/1890 - 05/21/1921
Ramblersville F.D.	
Ramblersville Pump & Hose	1900 - 1935
Richmond Hill F.D.	
Atlantic Hose	? - 08/05/1907
Clarenceville Ladder 2	00/00/1890 - 00/00/1913

Columbia Hose 1	00/00/1893 - 07/02/1907
Columbia Ladder 1	00/00/1893 - 07/02/1907
Morris Park Engine 1	11/11/1890 - 07/02/1907
Morris Park Ladder 1	00/00/1901 - 07/02/1907
Morris Park Hose	00/00/1890 - 07/02/1907
Pecare Hose 2	00/00/1890 - 07/02/1907
Richmond Hill Engine 1	? - 07/02/1907
Richmond Hill Ladder 1	? - 07/02/1907

Rockaway Beach (Oceanus) F.D.	
Atlantic Engine 1	03/01/1887 - 09/01/1905
Atlantic Hose 1	03/24/1890 - 09/01/1905
Oceanus Ladder 1	10/13/1886 - 09/01/1905
Remsen Hose & Chemical 3	01/28/1895 - 09/01/1905
Rescue 3	? - 09/01/1905
Sam Meyers Ladder 2	03/20/1897 - 09/01/1905
Seaside Engine 1	02/13/1895 - 09/01/1905
Seaside Hose 1	01/13/1888 - 09/01/1905
Volunteer Hose 1	00/00/1887 - 00/00/1890
Volunteer Hose 2	10/22/1887 - 09/01/1905

Rockaway Point F.D.	1916 - Active

Rosedale F.D.	
Rosedale Chemical Engine	00/00/1909 - 09/16/1927
Rosedale Ladder 1	10/00/1904 - 09/16/1927

Roxbury F.D.	1910 - Active

St. Albans F.D.	
St. Albans Ladder 1	00/00/1916 - 07/16/1921

Springfield F.D.	
Chemical 1	00/00/1897 - 09/16/1927
Ever Ready Ladder	00/00/1897 - 09/16/1927
Excelsior Hose	00/00/1903 - 09/16/1927
Springfield Engine 1	00/00/1897 - 09/16/1927
Springfield Hose 1	00/00/1897 - 09/16/1927

West Hamilton Beach F.D.	1928 - Active

Whitestone F.D.	
Columbia Hose 1	09/07/1891 - 10/01/1915
Indian Hose 2	10/06/1894 - 10/01/1915
Whitestone Engine 1	09/28/1887 - 10/01/1915
Whitestone Hose 1	00/00/1889 - 00/00/1908
Whitestone Ladder 1	07/21/1871 - 10/01/1915

Woodhaven F.D.	
Americus Hose 3	09/23/1896 - 09/01/1915
Americus Ladder 1	09/07/1888 - 09/01/1915
Aqueduct Hose 4	08/29/1902 - 09/01/1915
Deraismes Engine 1	05/27/1901 - 09/01/1915
Deraismes Hose 1	02/29/1892 - 09/01/1915
Jones Ladder 3	01/01/1899 - 09/01/1915
Pleasant Point Pump	? - 09/01/1915
Point Pleasant Ladder 1	? - 05/281913
Southside Engine & Pump	00/00/1897 - 09/01/1915
Union Course Ladder	01/23/1892 - 09/01/1915
Woodhaven Hose 1	00/00/1889 - 09/01/1915

STATEN ISLAND

Charleston/Kreischerville F.D.	
Washington H&L 1	00/00/1894 - 10/01/1905

Dongan Hills F.D. (originally part of the Edgewater F.D.)	
Cromwell Engine 8	00/00/1903 - 06/01/1912
Cromwell Hose 8	00/00/1906 - 06/01/1912

Edgewater F.D.	
Atlantic Hose 1	00/00/1859 - ?
Benjamin Brown Hose 3	06/02/1869 - 11/01/1905
Chemical Engine 1	00/00/1903 - 11/01/1905
Clifton Engine 8	06/02/1863 - 00/00/1893
Clifton Hose 6	09/16/1863 - 00/00/1893
Columbia H&L 5	03/15/1880 - 11/01/1905
Daniel Tompkins Engine 5	00/00/1857 - 00/00/1867
Enterprise H&L 1	12/05/1856 - 11/01/1905
Excelsior Bucket 1	10/20/1858 - 00/00/1890
Excelsior Hose 1	00/00/1890 - 11/01/1905
Grasmere & Concord Hose 10	02/20/1893 - 10/01/1905
Liberty Hose 3	circa 1865
Neptune Hose 1	12/15/1869 - 03/16/1878
	03/16/1878 - 11/01/1905
Neptune Engine 6	04/12/1857 - 11/01/1905
Niagara Engine 5	09/12/1867 - 11/01/1905
Peter Wiedner Hose 11	00/00/1890 - 11/01/1905
Protection Engine 7	00/00/1858 -11/01/1905
Protection Hose 7	1869 -
Red Jacket Hose 13	00/00/1902 - 11/01/1905
Relief Bucket 2	05/11/1863 - 00/00/1884
Relief Hose 2	00/00/1890 -11/01/1905
Rescue Engine 9	05/01/1879 - 11/01/1905
Robinson Hose 9	02/17/1880 - 11/01/1905
Rosebank Hose 14 (12)	? - 11/01/1905
Scott Hose 7	11/00/1890 - 11/01/1905
South Beach Hose 12	00/00/1890 - 11/01/1905
Tiernan H&L 6	07/00/1893 - 11/01/1905
Tompkinsville Fire Police	00/00/1859 - 00/00/1893
United Hose 10	? - 11/01/1905
Washington H&L 1	00/00/1894 - 11/01/1937
Wadsworth Hose 4	03/08/1890 - 11/01/1905

Great Kills F.D.	
Constitution H&L 1	00/00/1890 - 12/20/1928
Great Kills Engine 1	00/00/1890 - 12/20/1928

Huguenot F.D.	
Huguenot Engine 1	00/00/1895 - 11/01/1937

Midland Beach F.D.	1917 - 1919
Midland Beach Fire Company	

New Brighton F.D.	
Engine 1	
Hose 1	
Hose 4	

New Dorp F.D.	
Excelsior H&L & Engine Co.	00/00/1860 - 00/00/1894
Excelsior Engine 1	00/00/1894 - 00/00/1915
Excelsior H&L 1	00/00/1894 - 00/00/1915

North Shore F.D.	
Active Hose 7	11/00/1894 - 10/01/1905
Alert Hose 1	00/00/1885 - 10/01/1905
Aquehonga H&L 1	01/01/1879 - 10/01/1905
Castleton Fire Patrol 1	00/00/1893 - 10/01/1905
Cataract Engine 2	08/19/1844 - 10/01/1905
Chemical Engine 6	12/301903 - 10/01/1905
Cherokee Engine Co.	~ 1900 - ?
Defender H&L 7	12/30/1900 - 00/00/1937
Ellsworth Hose 5	00/00/1862 - <1874
Empire H&L 1	00/00/1891 - ?
Friendship Hose 2	05/16/1860 - <1871
Friendship H&L 4	08/08/1876 - 00/00/1885
Granite H&L 2	08/04/1881 - 10/01/1905
Lafayette Hose 3	06/00/1885 - 10/01/1905
Medora H&L 3	06/10/1868 - 10/01/1905
Monroe Engine 5	09/11/1894 - 10/01/1905
New Brighton Engine 4	10/14/1856 - 10/01/1905
Oceanic H&L 6	04/17/1880 - Active
Port Richmond Engine 3	08/24/1859 - 01/01/1913
Port Richmond Fire Grenade Co.	1888
Richmond H&L 4	06/00/1895 - 10/01/1905
Richmond Engine 1	00/00/1901 - Active
Rocky Hose 9	~ 1903 - 10/01/1905
Starin Hose 5	09/00/1885 - 10/01/1905
Steady Stream Hose Co.2	06/05/1885 - 00/00/1904
Tompkins Hose 6	05/11/1890 - 10/01/1905
Union Hose 8	00/00/1902 - 10/01/1905
Washington Engine 1	10/07/1852 - 10/01/1905
Wyandote H&L 5	12/13/1886 - 10/01/1905
Zephyr Hose 4	02/22/1861 - 10/01/1905

Pleasant Plains Princes Bay F.D.

Amicita H&L 1	00/00/1900 - 11/01/1937
Citizen Engine 1	00/00/1900 - 11/01/1937

Rossville F.D.

Advance H&L 1	00/00/1903 - 11/01/1937

Stapleton F.D.

Excelsior (Bucket) Engine 1	10/20/1858 - 00/00/1875
Enterprise H&L 1	12/05/1856 - 00/00/1875
Protection Engine 3	10/11/1858 - 00/00/1875
Relief Bucket 2	05/11/1863 - 00/00/1875

Staten Island F.D. 1860 - 1869/70

Since the companies of the SIFD remained in service after the department was disbanded, each company is listed under the department it became a part of.

Totenville F.D.

Eureka Engine 2	07/03/1886 - 11/01/1905
Eureka H&L 3	~1890 - 11/01/1905
Protection H&L 1	00/00/1880 - 11/01/1905
Constitution H&L 1	00/00/1899 - 11/01/1905
Vigilant Hose 1	? - 1880

Travis F.D.

Oceanic H&L 1	00/00/1905 - Active

BRONX

City Island F.D.

City Island Ladder 2	? - 08/01/1899
Minniford Engine	? - 08/01/1899
Minniford Hose	? - 08/01/1899

Clason Point (Higgs Beach Vol. Fire Co.) F.D.

Aviation Company 3	00/00/1922 - Active
Higgs Beach Fire Company 1	00/00/1918 - 00/00/1946
Higgs Beach Fire Company 2	00/00/1920 - 00/00/1946

Eastchester F.D.

Defender Hose 1	? - 06/18/1923
Reliance Ladder 1	?
Washington Engine 1	?

Edgewater (Edgewater Park, Edgewater Camp) F.D.

Edgewater Volunteer Hose 1	01/00/1923 - Active

Fordham F.D. ? - 1874

Hunts Point F.D. ? - 1874

Kingsbridge F.D. ? - 1874

Marble Hill F.D.

Marble Hill Hose 1	05/18/1895 - ?

Melrose F.D. (Merged to form Morrisania F.D. 01/12/1856)

Lady Washington Engine 1	00/00/1851 - 01/01/1874
Protection Engine Company	00/00/1852 - 01/01/1874
Vigilant Hose Company	? - 01/01/1874

Morrisania F.D. (Organized 01/12/1856)

American Ladder 1	? - 01/01/1874
Cataract (Highbridge) Engine 3	01/00/1867 - 00/00/1868
Fleetwood Engine 3	00/00/1871 - 01/01/1874
Hope Hose 2	? - 01/01/1874
Hopkins Ladder 3	? - 01/01/1874
Jackson Engine 4	<1866 - 01/01/1874
Jordon L. Mott Ladder 2	? - 01/01/1874
Lady Washington Engine 1	00/00/1851 - 01/01/1874
Monitor Engine 2	04/00/1862 - 01/01/1874
Morrisania Ladder 4	? - 01/01/1874
Protection (*Honey Bee*) Engine 5	00/00/1852 - 01/01/1874
Vigilant Hose 1	? - 01/01/1874

Mott Haven F.D.(Merged to form Morrisania F.D. 01/12/1856)

Jackson Engine Company	? - 01/01/1874
Jordon L. Mott Ladder Company	? - 01/01/1874

Riverdale F.D.

Neptune Engine 3	? - 08/01/1884

Silver Beach (Bronx Beach) F.D. 1920 - 1971

Spuyten Duyvel F.D. ? - 01/01/1874

Tremont F.D.

Mutual Protection Bucket	? - 01/01/1874
Tremont Ladder	? - 01/01/1874

Unionport F.D.

Columbia Hose Company	? - 05/01/1896

Van Nest F.D.

Van Nest Engine 1	09/00/1906 - 05/01/1910
Van Nest Hose 1	? - 05/01/1910

Wakefield F.D.

Americus Ladder	? - 02/01/1896
Empire Engine 1	? - 02/01/1896
George Washington Hose	? - 02/01/1896
Laconia Hose	? - 02/01/1896
Nereid Engine 1	08/13/1890 - 04/24/1896

Washingtonville F.D.

Empire Engine 1	?

West Farms F.D.

Alert Engine 2	? - 01/01/1874
West Farms Engine 1	00/00/1851 - 01/01/1874

Westchester F.D.

Western Engine 1	?
Westchester Hose	? - 02/01/1896

Williamsbridge F.D.

Pioneer Ladder 1	? - 02/01/1896
Williamsbridge Engine 1	? - 02/01/1896

Woodlawn Heights F.D. ? - 01/01/1874

Index

M

N

O

P

Q

R

S

T

U

V

W

Y

Z